AF716284

THE SUBLIME INVENTION: BALLOONING IN EUROPE, 1783–1820

THE ENLIGHTENMENT WORLD: POLITICAL AND INTELLECTUAL HISTORY OF THE LONG EIGHTEENTH CENTURY

Series Editor: *Michael T. Davis*
Series Co-Editors: *Jack Fruchtman, Jr*
Iain McCalman
Paul Pickering
Advisory Editor: *Hideo Tanaka*

TITLES IN THIS SERIES

1 Harlequin Empire: Race, Ethnicity and the Drama of the Popular Enlightenment
David Worrall

2 The Cosmopolitan Ideal in the Age of Revolution and Reaction, 1776–1832
Michael Scrivener

3 Writing the Empire: Robert Southey and Romantic Colonialism
Carol Bolton

4 Adam Ferguson: History, Progress and Human Nature
Eugene Heath and Vincenzo Merolle (eds)

5 Charlotte Smith in British Romanticism
Jacqueline Labbe (ed.)

6 The Scottish People and the French Revolution
Bob Harris

7 The English Deists: Studies in Early Enlightenment
Wayne Hudson

8 Adam Ferguson: Philosophy, Politics and Society
Eugene Heath and Vincenzo Merolle (eds)

9 Rhyming Reason: The Poetry of Romantic-Era Psychologists
Michelle Faubert

10 Liberating Medicine, 1720–1835
Tristanne Connolly and Steve Clark (eds)

11 John Thelwall: Radical Romantic and Acquitted Felon
Steve Poole (ed.)

12 The Evolution of Sympathy in the Long Eighteenth Century
Jonathan Lamb

13 Enlightenment and Modernity: The English Deists and Reform
Wayne Hudson

14 William Wickham, Master Spy: The Secret War against the French Revolution
Michael Durey

15 The Edinburgh Review in the Literary Culture of Romantic Britain: Mammoth and Megalonyx
William Christie

16 Montesquieu and England: Enlightened Exchanges, 1689–1755
Ursula Haskins Gonthier

Forthcoming Titles

The Language of Whiggism: Liberty and Patriotism, 1802–1830
Kathryn Chittick

Romantic Localities: Europe Writes Place
Christoph Bode and Jacqueline Labbe (eds)

William Godwin and the Theatre
David O'Shaughnessy

Ebenezer Hazard, Jeremy Belknap and the American Revolution
Russell M. Lawson

The Spirit of the Union: Popular Politics in Scotland
Gordon Pentland

British Visions of America, 1775–1820: Republican Realities
Emma Vincent Macleod

THE SUBLIME INVENTION: BALLOONING IN EUROPE, 1783–1820

BY

Michael R. Lynn

Routledge
Taylor & Francis Group
LONDON AND NEW YORK

First published 2010 by Pickering & Chatto (Publishers) Limited

Published 2016 by Routledge
2 Park Square, Milton Park, Abingdon, Oxfordshire OX14 4RN
711 Third Avenue, New York, NY 10017, USA

First issued in paperback 2015

Routledge is an imprint of the Taylor & Francis Group, an informa business

© Taylor & Francis 2010
© Michael R. Lynn 2010

All rights reserved, including those of translation into foreign languages. No part of this book may be reprinted or reproduced or utilised in any form or by any electronic, mechanical, or other means, now known or hereafter invented, including photocopying and recording, or in any information storage or retrieval system, without permission in writing from the publishers.

Notice:
Product or corporate names may be trademarks or registered trademarks, and are used only for identification and explanation without intent to infringe.

BRITISH LIBRARY CATALOGUING IN PUBLICATION DATA

Lynn, Michael R.
The sublime invention: ballooning in Europe, 1783–1820. – (The Enlightenment world) 1. Science – Social aspects – Europe – History – 18th century. 2. Science – Social aspects – Europe – History – 19th century. 3. Ballooning – Europe – History – 18th century. 4. Ballooning – Europe – History – 19th century.
I. Title II. Series
303.4'83'094'09033-dc22

ISBN-13: 978-1-138-66121-9 (pbk)
ISBN-13: 978-1-8489-3016-2 (hbk)

Typeset by Pickering & Chatto (Publishers) Limited

CONTENTS

ACKNOWLEDGEMENTS

The immense popularity of ballooning at the end of the eighteenth century caught my attention and initially drew me to this project. Since starting my research, however, I have been reminded constantly of the continuing popularity of balloons. Whether seeing balloon prints at the bouquinistes along the Seine, dining on Montgolfière de Saint-Jacques at a restaurant on the Place des Vosges, or gratefully receiving another balloon-themed gift from friends and family, the importance of eighteenth-century aeronautics resonates and echoes across the centuries. Clearly, the legacy of the Montgolfier brothers continues unabated.

I have many people to thank for helping me with this project. The Research Scholar Program at Agnes Scott College funded the work of several undergraduate research assistants. Elizabeth Ott and Kerry Baxter each spent a summer examining aeronautics in England and America respectively. Michelle Beer sacrificed two years to this project and now knows more about ballooning in France and England than she ever would have thought possible and certainly would ever have wanted.

Archivists and librarians at a number of locations have been generous in getting me materials. Dr. Mary Elizabeth Ruwell, Trudy Pollok, and John Beardsley of the Department of Special Collections at the United States Air Force Academy Library in Colorado Springs were especially giving with their time and help. The librarians and archivists at the Centre de Documentation at the Museé de l'Air et de l'Espace, as well as at the Bibliothèque de l'Institut, the Archives de l'Académie des Sciences, the Bibliothèque Historique de la Ville de Paris, and the Bibliothèque Nationale all proved enormously accommodating. Stacy Schmitt and Debbie Adams in the Inter-Library Loan department at McCain Library at Agnes Scott College worked diligently to fill my many obscure requests. K. R. Johnson, head of the Library at Purdue University North Central also helped me resolve many last minute problems. Ginny Borolov made life as department chair manageable which allowed me to spend at least part of my days on this book. I also received assistance from Rex Morrow, Dean of the College of Liberal Arts at Purdue University North Central, who worked with me to find the time to finish this project. Mark Pollard at Pickering & Chatto provided excellent help and advice throughout the completion of the book.

A number of scholars have read and commented on parts of this project over the years; others have shared with me their own insights as well as their work in progress. I would like to thank Tom Broman, Mary Cain, Shelly Costa, Suzanne Desan, Jan Golinski, Mimi Kim, Tracey Laird, Paul Maravelas, Alexandre Métraux, David Reid, J.B. Shank, Larry Stewart, Marie Thébaud-Sorger, Alice Walters, and Simon Werrett. An earlier version of chapter five appeared as "Consumerism and the Rise of Balloons in Europe at the End of the Eighteenth Century" [*Science in Context* 21 (2008): 73-98].

Many people offered moral support during the years in which I worked on this project. My thanks go out to Mary Cain, Kristian Blaich, Jim Abbot, Chris De Pree, Lili Harvey, Karen Thompson, Shu-chin Wu, Cathy Scott, Gus Cochran, David Behan, Lesley Coia, and Julia Knowlton. My family, John and Sally Lynn, Kate Lynn and Lori Mehl, have also been very encouraging of my work. As always, Judy Lynn enthusiastically listened as I told her yet another story about a balloonist falling into the ocean, a globe that caught fire, or a riot that took place. She remains my staunchest fan and my kindest critic; without her I could never have written this book. Over the last ten years and especially during the last year of this project, when I was also transitioning to a new job and moving cross-country, Brandon and Tracey Laird illustrated the true meaning of friendship. Trips to the mountains in north Georgia, Friday-night games, karaoke, large quantities of beer, music, meals, movies and constant camaraderie all helped make my time in Decatur worthwhile and my work on this book easier. I dedicate this book to them.

For Tracey and Brandon Laird

LIST OF FIGURES

INTRODUCTION

The last decades of the eighteenth century witnessed an astonishing flurry of activity in the skies as savants and amateurs alike launched balloons into the air with great abandon much to the delight of the enormous audiences that gathered to watch and applaud their efforts. James Dinwiddie, for example, released a balloon with 'a peculiar form' near Buckingham Gate in London in late 1783 which resulted in an 'undulating serpentine motion'. This 'afforded great entertainment to the spectators'.[1] Mr Jackson, 'of the Hutton Rugby School, near Stokesley', set forth 'an elegant aerostatic globe' from the Market Place in Stockton in Durham, England in June 1784. The flight lasted only seven minutes but it ascended 'in a very pleasing manner the whole time it was visible'.[2] On 13 March 1785 Sieur Lhomond launched a globe from the Jardin des Tuileries in Paris. Three other balloons preceded the main globe; one represented the sun, the second golden thunder, while the third was without decoration but in which Lhomond experimenting with a mixture of hot air and hydrogen. The latter balloon exploded after take-off.[3] Throughout the second half of 1784 and into 1785 Monsieur Cailleau, from the town of Hauterive, released a series of small balloons. He and his assistant, Monsieur Rivière, used the Montgolfier balloon as their model.[4] In Florence the aeronaut Giard ascended in a balloon on 1 October 1811. He quickly rose to a great height and 'lost sight of the earth' leaving him 'numbed by the extreme cold, and himself nearly overpowered by sleep'. He eventually landed after nine hours aloft.[5]

These ascensions are not among the more famous typically described in books and articles about the genesis of ballooning. Most accounts of the origins of aeronautics begin with the moment when the Montgolfier brothers, Etienne and Joseph, released their hot-air balloon from the Place des Cordeliers in Annonay, France in 1783. The Montgolfiers chose their moment carefully, but not because of any scientific or technological factors. Instead, they picked that day and time because of the meeting of the local *Etats particulières*, a diocesan assembly in Annonay for their annual meeting. This group, composed of many local notables, acted as witnesses for the flight and attested to the efforts of the Montgolfiers in developing this new technology. They subsequently signed a document describing the launch which the Montgolfiers sent to King Louis XVI and the members of the French Royal Academy of Sciences. In the end, the Montgolfiers hoped the state would reward them handsomely for their efforts and acknowledge their place among the scientific elite.

From the perspective of lauding the moment when a new invention is first displayed before the public, historians are right to select this event to initiate a conversation about ballooning. However, the focus usually remains on the first successful voyage. While this is, of course, important, the flight ultimately looked incredibly similar to most other ascensions: someone scheduled the launch, a crowd gathered, the balloon took off, it wandered away, and eventually landed and, perhaps, caught fire. The significance of the Montgolfiers' efforts was not their launch but the precedent it established. From this moment on, launches remained almost entirely public affairs and witnesses – often paying for the privilege – attended in droves. Aeronauts could not steer their balloons and experiments often ended with the destruction of the globe; even once manned balloons appeared and aeronauts could control their descent to a greater degree, the landing was often dangerous and at least partly a controlled crash. Many of these aeronauts hoped for recognition and rewards for their efforts, just like the Montgolfiers. Thus, while the Montgolfier brothers deserve praise and have often been the focus of attention in histories of aeronautics, this study explores instead the broader phenomenon of ballooning. As the driving force behind the first public launch, the Montgolfiers are important; but any analysis of the development of an age fascinated by balloons requires that we quickly leave them behind and examine the many other balloonists, and the hundreds of other launches, across Europe and North America who brought this new invention to the forefront of the cultural lives of people at the end of the eighteenth century.

One aspect of ballooning that appears common throughout this period was the general enthusiasm people felt for it. Aeronautics was the wonder of the age and people compared balloonists to other famous explorers, mythological and real, including Vasco de Gama, James Cook, and Christopher Columbus.[6] Simon-Nicolas-Henri Linguet suggested that while Jason and the Argonauts had sought to recover the Golden Fleece, the 'modern Argonauts devote themselves only to the progress of the sciences, [and] to the perfecting of a very honorable discovery'.[7] Supporters described balloonists as pioneers and treated them as such although their points of comparison varied widely. Some people involved in this enterprise probably did not like it when people associated them with Franz Anton Mesmer, a man at the center of some debate, or Comus, a scientific popularizer noted for his theatrical demonstrations. Similarly, comparing the invention of balloons to inoculation, a medical treatment seen in askance, probably did not always assist the cause of aeronautics.[8] However, relating them to electricity and the discovery of a method to determine longitude certainly did help.[9] 'Never', claimed the author of the *Correspondance littéraire*, 'has a soap bubble occupied more seriously a troupe of children', than had the 'aerostatic globe'.[10] More positively, the distinctly partisan *Journal de Paris* claimed 'the success of aerostatic experiments in the capital [Paris], has produced a universal enthusiasm and all the cities in the kingdom have enjoyed this beautiful spectacle'.[11]

*****

Ballooning, like the Enlightenment, traversed Europe although historians have not always treated it so. In addition, ballooning initiated a massive cultural phenomenon that historians have rarely acknowledged beyond a few anecdotes. A full understanding of the importance of science during the age of the Enlightenment and Atlantic revolutions requires an explanation of how and why ballooning entered and stayed in the public consciousness. Spectacular by nature, ballooning has inspired historians to tell a few stories, but has not come under the intense scrutiny it deserves. Thus, rather than simply narrating a chronology of discovery, this book provides a cultural and social analysis of ballooning over the first quarter century after their invention. During this period, a paradox emerged that partially defined the cultural importance of this invention. On the one hand, a mass culture emerged surrounding ballooning. Aeronauts captured the hearts and imagination of the entire continent, and beyond, and balloons became a symbol of Enlightenment, state power, and scientific progress. People of all social levels, educational backgrounds, genders, and scientific ability lauded balloons. On the other hand, balloons completely failed to fulfill their potential during this period. A method of steering balloons never materialized thus transforming what might have been an exceptionally powerful tool into mere entertainment.

Books relating this story abound, starting with Tiberius Cavallo's *The History and Practice of Aerostation* which appeared in 1784.[12] While Cavallo wrote his history of aeronautics at a somewhat early stage, a number of other authors offered versions during the nineteenth century including Gaston Tissandier and Monck Mason, both of whom wrote narratives describing its development from the Montgolfiers to their own times.[13] A twentieth century account, written by L. T. C. Rolt, provides a brief, and broad, survey of ballooning from 1783 to 1903, the year the Wright brothers made their first flight near Kitty Hawk, North Carolina. Rolt's book enthusiastically relates many great stories about aeronautics.[14] A number of other broad surveys on this subject have also appeared in a variety of languages.[15] One of the best books on this subject is Charles C. Gillispie's monograph, *The Montgolfier Brothers and the Invention of Aviation, 1783-1784* which appeared, along with a few other books in France, around the bicentennial of the invention of ballooning.[16] Gillispie's book, based on intensive research into the Montgolfier family papers and the documents in the Fonds Montgolfiers collected at the Musée de l'Air et de l'Espace, explores the early history of ballooning with a strong concentration on Etienne and Joseph Montgolfier. Other aeronauts appear in this study, but mostly as showmen who distract from the real importance of ballooning. In addition, Gillispie focuses his attention largely on the first two years of ballooning, discusses France almost exclusively, and really uses ballooning as a starting point for a larger conversation centered on nineteenth-century technological advances in steam engines and railways. Marie Thébaud-Sorger's recent contribution to this literature, *L'Aérostation au temps des Lumières*, also focuses on the first few years of aeronautics although she examines the broader cultural environment throughout France.[17]

Some authors tackle the subject through a national or regional approach. Tom D. Crouch's *The Eagle Aloft: Two Centuries of the Balloon in America* examines the North American case. However, in treating two hundred years in one volume, Crouch necessarily gives the earlier period short shrift.[18] J.E. Hodgson's tome, *The History of Aeronautics in Great Britain* provides a great national narrative, and an excellent collection of stories about various balloonists and their launches.[19] Davide Arecco's *Mongolfiere, science et lumi nel tardo settecento* begins with an account of the origins of aeronautics in France before turning to a case by case presentation of the efforts of various Italian balloonists in the eighteenth century.[20] John Penny's *Up, Up and Away: An Account of Ballooning in and around Bristol and Bath, 1784-1999* has a fairly narrow regional focus although it is chronologically very broad.[21] While some historians do offer some comparative analysis of ballooning in different locations most tend to treat each area as independent.[22]

Some balloonists have been the focus of biographical studies. The Montgolfier brothers, of course, have received much attention and even some general studies of aeronautics focus considerable attention on them. However, aeronauts such as Jean-Pierre Blanchard, Jean-François Pilâtre de Rozier, Vincenzo Lunardi, Wilhelmine Reichard, and James Tytler have also served as the subjects of books and other scholarly efforts.[23] So many historians have devoted attention to aeronautics that in the first half of the twentieth century several bibliographies were published to help researchers.[24] In contrast, this study uses a broad geographical but relatively narrow chronological approach that does not favor any particular balloonist.

*****

Although people had long imagined flight, hot-air and hydrogen balloons were new and it is this newness that makes them particularly fascinating. Aeronauts and their audience felt free to use and appropriate this invention in any way they wished. Historians have often studied changes in attitudes toward ideas or objects; but in this case, balloons represented a tabula rasa on to which people could form their own vision. Thus, people applied balloons to existing cultural forms – including novels and poems, furniture and clothes – but they also developed different venues in which balloons appeared; the most important of these new arenas consisted of the events surrounding a commercial launch. As a new object, entrepreneurs had to develop ways in which they could sell them, the state had to decide how to control something that was potentially dangerous, and savants needed to discover ways to use them. Examining how and why people appropriated this invention into their lives illuminates some aspects of their cultural mentalities. In this way, balloons are useful as objects onto which people can place their values and ideas.[25]

This book explores the 'universal enthusiasm' for balloons from the inception of aeronautics in 1783 until about 1820. The starting date is somewhat clear as it stems from the first flight of a lighter-than-air object. However, it can be seen as problematic. Other savants besides the Montgolfier brothers had been experimenting with using gases to fill globes and create floating objects. In addition, there were efforts by some individuals to develop mechanical craft that could fly through the air. A rich, literary tradition also existed in which human flight, through gases, machines, or other means, appeared as a possibility. From this perspective, the starting point of 1783 can be viewed as somewhat arbitrary. However, this book's focus centers on flights and their impact on the culture of the times; as such, the moment when people first became aware of the existence of working aerostatic globes marks the logical beginning for this study. This study ends in approximately 1820. This period marks a shift in the ways in which balloonists inflated globes. At that time, Charles Green began using coal gas to fill balloons. Coal gas proved much cheaper compared to hydrogen although it had less lifting power. In addition, in London they began laying down gas mains for homes and businesses. This meant that an aeronaut could hook up their balloon to a gas main and fill it fairly rapidly and easily. Thus, the move toward coal gas changed the nature of aeronautics by lowering the cost and making it even easier to fill and launch a balloon.[26] This shift did not happen universally but it does mark an important change. Thus, this study ends at about the same time innovators introduce coal gas into the field of aeronautics. In addition to this technological change, the period of time under consideration – about thirty-five years – provides a decent snapshot of the field. Most historians focus on either the first few years or explore the topic from the beginning up to the time they are writing. Thus, an examination covering the first few decades will provide an alternative view of how ballooning entered and stayed in the public consciousness.

Ballooning's international stature both as science and as entertainment, combined with the propensity of some aeronauts to travel around Europe and North America, encourages a broad geographical approach to this study. Just as with ideas and goods, aeronautics circulated freely and ignored national borders. The spread of ideas about and knowledge of ballooning is inseparable from the spread of actual balloons.[27] Thus, this study tends to cross from one state to the next, much as did aeronauts. In addition, this book examines how and why balloons came to occupy such a strong place within the culture of late eighteenth and early nineteenth century Europe and North America. Certainly, it might be enough to say that humans had conquered nature and to point out that a balloon soaring through the sky presents an awesome sight. However, for people not formerly inundated with examples of flight, and for whom a balloon was a wholly new notion and object, aeronauts and their supporters needed to exert some effort to demonstrate to people why they should attend launches and spend their money on such an ephemeral event.[28] Aeronauts only had to change their tactics once it became clear that people could not steer balloons and that this invention, as yet, served no useful purpose other than to entertain. The focus then shifted to include, sometimes entirely, entertainment. Even there balloonists had to sell their product, and all of the associated paraphernalia, and transform aeronautics into a desirable commodity.

Balloons came to occupy the hearts and minds of a wide variety of people. Men and women from the nobility to the peasantry all had the chance to view launches. Savants and amateurs alike felt they had something to say about aeronautics; the fact that individuals without scientific training conducted successful launches meant that scientific academies or their representatives remained unable to dominate this new field of study even if they had wanted to do so. At the same time, the funding of aeronautics, through ticket sales and subscriptions, meant the general public had much greater influence over the practice of ballooning than savants would have liked. Many governments, from nations to cities, felt the need to try and control this new invention and the people who thronged to see it even as they readily relegated the funding of experiments to the general public. The broad spectrum of people involved in aeronautics, as balloonists or witnesses, authors or merchants selling balloon-wares, necessitates a similarly broad approach to understanding the importance of this new invention in the culture of the period. It also necessitates a broad use of sources including newspaper accounts and advertisements along with memoirs and letters and scientific treatises.[29] Balloons became important intellectually, socially, politically, economically, and culturally; thus, this study seeks to understand balloons at each of these levels.

******

This book is divided into six substantive chapters. Chapter 1 offers an introduction to the history of flight in Europe. Starting with ideas of flight prior to 1783, the chapter explores the spread of ballooning across Europe and North America during the first years after its invention. The traditional narrative concentrates on France and England but, as this chapter demonstrates, ballooning spread much wider than historians usually acknowledge. Additional examinations of disasters and accidents, along with the invention of parachuting, help illuminate why aeronauts continued to keep the public's attention and the ways in which balloonists maintained their place within the culture of the day.

The second chapter discusses the relationship between balloons and Enlightenment. From the moment of their invention, balloons became symbolic of the enlightened age, the ability of mankind to conquer nature, and the ultimate utility of science to society. Commentators suggested a number of areas that might benefit from balloons: these included science, transportation, and commerce. More fancifully, some suggested that balloons might help thieves and lovers escape. Problems emerged, however, over exactly how balloons might be useful, especially since no one could devise a method for steering them. Thus, this chapter scrutinizes the problematic relationship between utility, Enlightenment, and aeronautics.

Chapter 3 analyses the social side of ballooning. This includes a discussion of the social position of balloonists as well as an exploration of who the audience for launches might have been. Aeronauts experienced difficulty controlling the area around their launches and how people viewed the balloons. This was particularly true once the balloon reached a certain height. This helps explain the importance of controlling access to the workshops where aeronauts manufactured their balloons as well as the desire of the local and state authorities to police the practice of ballooning. There is a strong element of elitism associated with ballooning that rarely gets mentioned in the existing literature; this elitism, in which peasants are isolated from this invention, contrasts significantly with the popular nature of aeronautics. How did savants try to maintain control of this invention and assert their special place in the intellectual hierarchy at the expense of amateurs?

The fourth chapter examines the place of ballooning within political culture. Almost immediately state authorities integrated balloons into royal, as well as republican and imperial, festivals. The military also targeted balloons as potentially useful. The French revolutionary government even created a Balloon Corps which played an important role at the Battle of Fleurus and accompanied Napoleon on his campaign to Egypt. This chapter will also focus on efforts by states to control launches, perceived by some leaders as dangerous politically (because of their potential use by spies) and as risky (as fire hazards). Failed launches also led to riots, something the state always preferred to avoid.

Chapters 5 and 6 explore, broadly speaking, the economics and culture of aeronautics. The fifth chapter examines the selling of ascensions. Although balloons failed to prove particularly useful for savants conducting experiments, as a form of entertainment they excelled. In addition, they arrived on the scene at exactly the same moment as the so-called 'consumer revolution'. While this revolution is debatable, it is certainly true that the eighteenth century experienced an expansion of commercial choices.[30] The late eighteenth and early nineteenth century saw the development of the consumer marketplace in Great Britain and, to varying degrees, in France, North America and elsewhere. How did balloons fit into this emerging market with the related growth in commercial venues? As a new product, balloons offer an excellent case study into questions of product innovation, the role of advertising, and national characteristics of a consumer society. Chapter six discusses the material culture surrounding the invention of balloons. Manufactured goods with balloon-related themes, such as plates, clocks, furniture, clothes, and so on, became enormously fashionable in the late eighteenth century. Literary and artistic culture more generally, in the form of broadsides, paintings, poems, novels, almanacs, and scientific treatises, produced enormous numbers of cultural artifacts related to aeronautics. The ways in which literary works discuss, and artistic works depict, the practice of ballooning help demonstrate the breadth of impact ballooning had on European culture.

1 THE EMERGENCE OF AERONAUTICS

Historians have recounted the story of ballooning many times, in many different ways, geared towards a variety of audiences. Most of these accounts take the long view and aim to move from 1783 to the present (or at least to the birth of modern aviation), often utilizing an international approach although occasionally concentrating on a particular country or region.[1] On the other hand, a number of historians have explored ballooning from a more biographical approach. While occasionally this means a study of figures such as Pilâtre de Rozier, Lunardi, Tytler or Garnerin, most often the subject of these studies are the Montgolfier brothers, Étienne and Joseph.[2] The historiography of ballooning, then, typically proposes its subject as an important invention and key precursor to modern flight, as an analysis of a particular nation's or region's involvement in this endeavour, or as an explication of some of those individuals involved in its early practice.

The version of this history here, on the other hand, neither attempts completeness nor limits itself to particular locations or people. While some chronological description appears, this narrative does not want simply to describe the flow of events. Instead, this chapter highlights certain themes in an effort to demonstrate the elements of ballooning that helped account for its popularity and suggest how and why aeronautics spread across Europe. In other words, rather than offer an internal account of the growth and development of ballooning as a particular type of technology, this chapter sketches an outline of ballooning based on what the general public at the end of the eighteenth century might have found important. Since the overall argument of this book hinges on popular attitudes toward ballooning, it becomes important to highlight those aspects of this new invention that dominated the public imagination. The first section provides a description of flight as conceived prior to the invention of balloons. This is followed by an analysis of the geography of ballooning activities and a discussion of the ways in which ballooning spread during the initial years of flight. The public always lauded the first people to fly in a particular region, the first to cross a particular barrier, the first women aeronauts, the highest or longest flights and so on. Enthusiasm for ballooning extolled, in part, the heroic aspect of flights. Ballooning, however, was also dangerous and the next section explores the public fascination with balloon-related accidents and deaths. Last, the appearance of parachuting is discussed, both as a response to the dangers of aeronautics as well as a method for enhancing the entertainment value of ballooning.

Preflight

> What's the news of the day,
> Good neighbour, I pray?
> They say the balloon
> Is gone up to the moon![3]

Tales of flight are as old as humanity. The gods of Greek and Roman mythology frequently flew, either in the guise of animals or through their own volition. Unsurprisingly, early Greek and Roman thinkers also turned their attention to the art of flying. Archytas, a Pythagorean philosopher and mathematician living at the time of Plato, reputedly crafted a wooden model of a dove that could fly.[4] From antiquity to the age of Enlightenment, numerous authors and savants posited the possibility of flight, sometimes in fictional form and other times with more serious suggestions for how to achieve air travel. Thus, in the decades and centuries before 1783, the general population already had a wealth of competing visions of flight.

Numerous authors speculated about the possibility of human flight in the period before the Montgolfier brothers demonstrated their invention.[5] These include flight as a peripheral notion in a larger work as well as books in which flight provided a convenient device for travelling to distant lands (or even to stars, moons and planets). Cyrano de Bergerac, for example, attached bottles of dew to himself; when heated, he claimed, the dew would vaporize and rise up. This allowed Cyrano to jump all the way to the sun and moon. Travelling that far enabled him to discuss social and philosophical problems from a geographical and critical distance.[6] Such fanciful descriptions abounded in the eighteenth century including Voltaire's *Micromégas* and less well-known books such as Joseph Galien's *L'Art de naviguer dans les airs* or Domingo Gonsales's *The Man in the Moone: Or, a Discourse of a Voyage Thither*.[7] Galien's book outlines an enormous, impossibly large, lighter-than-air craft designed to transport a military force into the middle of Africa. Gonsales, on the other hand, uses his machine as a means of escape and travel. The frontispiece illustrates his method; birds are harnessed to a frame with a sail on one end and a seat for the voyager.

Jean-Jacques Rousseau, in *Le Nouveau Dédale*, suggested the use of compressed air, judiciously released, would, along with a rudder, allow someone to fly through the air.[8] Restif de la Bretonne goes back to the idea of human wings in his book, *La Découverte australe*.[9] In his novel *The History of Rasselas*, Samuel Johnson included a brief chapter, 'A Dissertation on the Art of Flying', in which an artist labours to create wings for himself and then tries to use them to fly, only to fall in the water and nearly drown.[10] Earlier, in an issue of the *Rambler*, Johnson wrote about a savant who claimed to have 'twice dislocated my limbs, and once fractured my skull in essaying to fly'.[11] The list of authors who utilized some sort of flight could go on almost indefinitely and includes Jonathan Swift, Aphra Behn and Daniel Defoe, to name just a few.[12]

Writers of fiction received inspiration for their flights of fancy in the efforts of a number of innovators who tried to launch aircraft of various kinds in the period before 1783. Some savants explored the possibilities of flight with early versions of helicopters. Most famously, Leonardo da Vinci sketched a helicopter in his notebooks. Similarly, Athanasius Kircher worked on the idea of creating a flying machine.[13] These individuals belonged to a much longer tradition that continued up to and after the invention of ballooning. The seventeenth-century Italian savant Tito Livio Burattini left Venice for the Polish court to work for King Wladislas IV as a mathematician and Master of the Mint. While there, he developed designs for a 'flying dragon,' a heavier-than-air machine with a series of eight pairs of wings that flapped to provide life and forward motion.[14] Burattini's contemporary Chrsitaan Huygens heard news of Burattini's device and worked on a version of his own which utilized a sort of propeller for propulsion.[15] The Jesuit scholar Francesco Lana sketched a model for a flying vessel modelled on a sailboat. In 1709 a Portuguese savant named Bartholomeu Lourenço de Gusmao purportedly created and demonstrated a flying machine. This event, oft-reported after the advent of the ballooning age, remained unsubstantiated. David Bourgeois, for example, discusses it briefly in his 1784 book *Recherches sur l'art de voler*. Here Bourgeois lists every account of flying he could identify in the period before 1783. He starts with Dedalaus and Archytas but also includes Roger Bacon, Da Vinci, Kircher and Gusmao. In all he mentions over thirty people who engaged with the problem of flight.[16] Later, in 1784 the mechanic and scientific popularizer François Bienvenu along with his partner, a naturalist named Launoy, published a short prospectus announcing their flying machine, a sort of early helicopter.[17]

Prior to the invention of lighter-than-air flight, however, other savants struggled with success and lost. Jean-Pierre Blanchard developed what he called a 'vaisseau volant', or flying vessel based on one of his earlier inventions, a prototype for a bicycle called a velocipede.[18] He began touting his vessel as early as 1781 in an attempt to gain attention, notoriety and funding.[19] Although poorly educated, Blanchard exhibited enormous talents as a mechanic. The 'vaisseau volant' used foot pedals as well as hand levers to operate four wings. These wings flapped and, in theory, would elevate the operator who sat enclosed in a cockpit. The result, if it had worked, would have been something like a helicopter in design (see Figure 1.1). Blanchard, who claimed his machine could travel at speeds up to seventy-five miles per hour, announced a demonstration in May 1782.[20] He planned the experiment for a Sunday and sold tickets to interested individuals. After his first announcement Blanchard published a letter in the *Journal de Paris* in which he claimed to have received 'eleven or twelve hundred letters' about his experiment; this enthusiasm prompted him to add a second demonstration.[21] A few days later, however, Blanchard somewhat ruefully admitted he had to delay his spectacle for three weeks during which time he would 'perfect it'.[22]

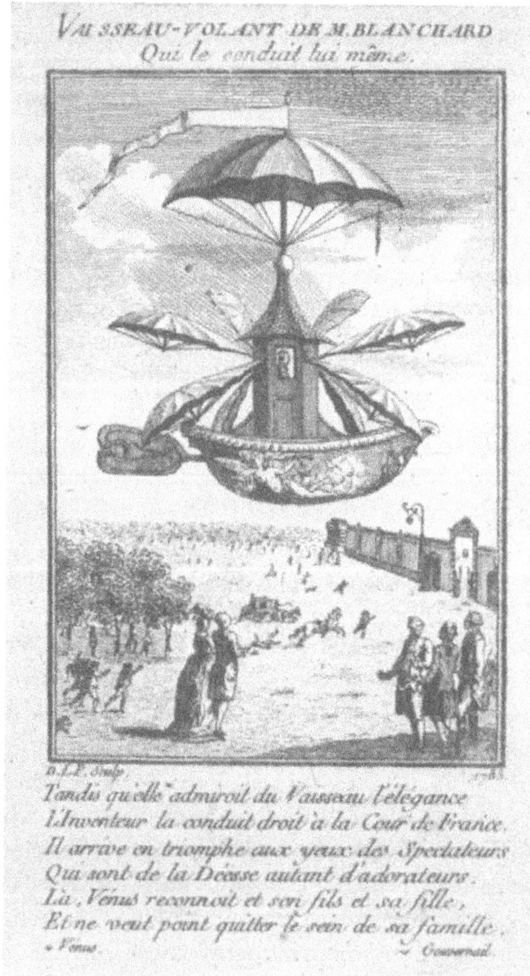

Figure 1.1

Meanwhile, some savants took the opportunity of critiquing Blanchard before he even tried to get his machine off the ground. The celebrated astronomer Jérome-Joseph Lalande, for example, compared Blanchard to the infamous dowser Barthelemy Bléton and scolded the editors of the *Journal de Paris*.[23] He claimed they discussed such absurdities so often that their readers might be convinced the editors believed in any and all nonsense. Later, during the French Revolution, Lalande softened his attitude toward Blanchard and participated in some balloon flights with him during which Lalande conducted meteorological experiments.

Other authors wrote more favourably about Blanchard's machine. An engineer and engraver named Martinet, for example, wrote a letter to the *Journal de Paris* in which he claimed Blanchard had performed 'a little experiment', and attested to Blanchard's efforts in order to 'arrest the cloud of sarcasms which cover it daily with jealousy and incredulity.'[24] Unfortunately, at least for Martinet and Blanchard, the 'vaisseau volant' was destined to stay on the ground. When the time came to finally stage a launch Blanchard claimed the rain interfered with the efficacy of his machine and instead chose to read a paper about the device.[25] Arguably Blanchard's machine had a fighting chance of getting off the ground; other attempts at flight appeared doomed right from the start. The Marquis de Bacqueville, for example, built a pair of wings and, in 1742, leapt off the roof of his house. Luckily, his home was located alongside the Seine River in Paris. This was ideal in his mind since, if his experiment failed, he would fall into the water. Unfortunately, when this eighteenth-century Icarus fell, he landed on the deck of a washer-woman's boat and broke his leg.[26]

The Advent and Geographical Expansion of Ballooning

A full discussion of every example prior to 1783 where an individual wrote a fictional account of flying or, perhaps, actually attempted to take to the skies, could occupy an entire book by itself. So, too, could a discussion of the origins and spread of ballooning; a number of scholars have undertaken to provide just such a narrative account of aeronautics. As noted above, however, it is not the purpose here to rewrite that narrative or to attempt to discuss in full every balloonist in Europe from 1783 to 1820. Instead, this section couples a brief description of the birth of the ballooning phenomenon with an analysis of the spread of this new science throughout Europe and the Americas.

Although, and not without justification, the Montgolfier brothers receive the lion's share of the attention from scholars narrating the origins of flight in Europe, the broader picture includes a larger cast of characters all working independently towards the same object. Joseph Black, for example, experimented with Henry Cavendish's new hydrogen gas. In particular, he tried to fill various bladders to see if they might float. The bladders, however, were either too porous or too heavy and his experiments failed. After the Montgolfier brothers launched their first balloon, Black did not argue for a prior claim to the invention although others, like Jean-Pierre Blanchard, did.[27] Some savants, like Joseph Priestley and Tiberius Cavallo, also experimented with hydrogen. Thus, the geography of the invention of ballooning extends beyond a single day in Annonay, France. Even the choice made by the Montgolfier brothers, to conduct their first public launch on 5 June 1783, had less to do with the timing of their invention than it did with the convenience of having the appropriate audience present for the experiment. The Montgolfiers had certainly launched balloons, in private, prior to 5 June. However, the moment of revelation was just as important as the moment of discovery and, in this case, Joseph and Etienne felt the necessity of an audience which included local notables so that their experiment might gain in status.[28]

As the moment and place of invention for ballooning really encapsulates a much larger period of time and place than simply 5 June 1783 in Annonay, France, so too the spread of ballooning cannot be traced in a linear fashion. Instead, the spread of ballooning needs to be seen as a function of the dissemination of information at the end of the Enlightenment, particularly through the periodical press, correspondence and treatises.[29] In addition, this new invention needs to be viewed alongside the ability of specific individuals to interpret what they read and recreate the globe. In other words, people in a specific area needed to have access to the right sources of information, and the mechanical know-how to construct a balloon, along with the willingness to risk their reputations, if not their lives, for the sake of aeronautics. Some authors attempted to track this movement across Europe. Tiberius Cavallo, for example, traced the rapid spread of aeronautics although his focus is largely on France and Great Britain.[30]

The first and easiest illustration of this sort of geographical movement comes with the spread of knowledge about balloons from Annonay to Paris. As the Montgolfier brothers intended, information about their experiment raced to the capital with great speed as news of their success travelled north. Famously, Jacques-Alexandre-César Charles and Jean-François Pilâtre de Rozier jumped on the aeronautic bandwagon and conducted experiments of their own, using hydrogen and hot air respectively. Similarly, on 11 September 1783 the Baron de Beaumanoir launched small balloons, a foot and a half in diameter, from the Hôtel de Surgoris in Paris; he gave a repeat performance later that same month.[31]

News of the invention spread as quickly, if less intentionally, to other parts of France. In Strasbourg, the savants Frédéric-Louis Ehrmann and Georg Christoph Würtz published a book on ballooning early in 1784 as did a local professor of mathematics, Christian Kramp.[32] Meanwhile, back in November 1783, local authorities in that city charged two mechanics named Degabriel and Pierre with the task of creating a balloon. They raised a subscription, set at six livres per ticket, and picked a launch date in May 1784. The state almost thwarted Degabriel and Pierre when they considered cancelling the event due to a fear of accidents. However, after proving to the satisfaction of the local magistrates that their balloon was safe, they attempted the experiment. Unfortunately, it did not go as planned. Bad weather forced delays and they had trouble inflating the globe. Another launch, planned for August, probably never took place.[33] A Piedmontese savant named Francesco Adorne performed his own balloon experiment along with a local man named de Winter in front of a 'large mob' on 15 May 1784. The balloon descended soon after it had taken off and caught fire, subsequently igniting a nearby building. Adorne attempted further experiments later in the same year.[34]

While the citizens of Strasbourg discussed balloons and attempted some launches, people in other cities around France grappled with this new invention with equal enthusiasm. In Lyons, the Montgolfier brothers along with Jean-François Pilâtre de Rozier and several local nobles released a balloon in January 1784; a second ascension took place in Lyons before the King of Sweden on 4 June 1784 during which the first female aeronaut, Madame Thible, went up. In Mâcon, a local 'avocat du roi' named Cellard de Castelais sent forth a globe on 15 February 1784. Twenty feet in diameter this balloon stayed aloft for two hours.[35] In Dijon, Louis-Bernard Guyton de Morveau and the Abbé Bertrand ascended in a balloon in April 1784 while by May, two local savants named Bonin and Maret had taken off from Marseilles. Later that same month Maret arose with a merchant named Brémond.[36] By June, Darbelet, a professor of natural philosophy, along with an architect named Chalfour and another local citizen named Desgranges, conducted a manned launch in Bordeaux while in Nantes the local lieutenant-general, Constard de Massy, and a professor of natural philosophy named Mouchet oversaw a flight lasting fifty-eight minutes. Two professors in Rodez, in Guyenne, ascended in a balloon on 6 August 1784 and stayed aloft thirty-five minutes.

Countless numbers of other cities in France also conducted launches of their own in 1784. Even the French colonies were quick to attempt experiments. In Cap François, Saint Domingue, now Haiti, local savants released a balloon on 31 March 1784; this proved to be the first aeronautical experiment in the Americas, even ahead of the United States. Several additional ascensions took place in the following months, much to the delight of women in particular, at least according to one commentator.[37] Thus, the distance, in terms of scientific and technological dissemination, from Annonay to Paris and then to Dijon, Lyons, Strasbourg, Cap François and Bordeaux was fairly slight with enough information passing in less than a year for individuals in those cities to launch balloons independently (except in the case of Lyons where the Montgolfier brothers arrived to help) based largely on descriptions received through newspaper articles and eyewitness accounts.

While having information move quickly across France is notable, even more impressive is the speed with which enough information moved across national borders allowing local people in Great Britain, Spain and on the Italian peninsula to conduct their own experiments. Two noblemen, named de Lamanon and Nappion, along with a doctor named Bonsvoiser, released a small, unmanned balloon in Turin on 13 December 1783.[38] Fransesco Henrion, an engineer and mineralogist from Pistoia, launched his globe early in 1784 from the Ponte alla Carria in Florence; this initiated a frenzy of ascensions in Tuscany including one on 1 February 1784 that 'rose to a great height' but 'caught fire and came down immediately.' This, along with other accidents, led the Grand Duke of Tuscany to issue an edict on 13 April 1784 forbidding such activities in his territory.[39] In Milan, Count Paulo Andreani and the Gerli brothers, Agostino and Carlo Giuseppe, conducted an experiment on 25 February 1784.[40] This event, funded by the nobleman Andreani who had developed an early interest in aeronautics, was managed by Agostino, an architect, and Carlo, a painter. They constructed the globe based on 'the good principles' developed by Andreani.[41] Tiberius Cavallo claimed Agostino and Carlo did much of the work and described them as 'persons of mechanical genius'.[42] The two brothers accompanied Andreani on the flight and waved 'happily to the spectators,' more than 2,000-strong, who showed up to watch.[43] The Gerli brothers subsequently published some of their views on how best to construct a balloon. In particular, they argued for the separation of the passenger car from the brazier used to maintain the temperature of the air in the balloon.[44]

Savants in the Venetian Republic followed suit in April 1784 when the Zanchi brothers launched a balloon. Ballooning became a very exciting topic of conversation on the Italian peninsula from an early date; local savants even created a journal, published in Milan, devoted entirely to aeronautics. The *Giornale aerostatico*, however, lasted only one year.[45] Other savants, such as Carlo Amoretti, published treatises on this new invention as early as 1783.[46] However, balloons did take longer to reach other parts of the Italian peninsula. Francesco Zambeccari, already famous for several ascensions in Great Britain, performed one the first flights in Bologna, but not until 1803. Similarly, Vincenzo Lunardi, also famous for flights in Great Britain, launched the first balloon in Naples, for the monarchs King Ferdinand and Queen Maria Carolina, in 1789 at a private experiment.[47]

The Spanish initiated their first balloon experiments early in 1784. Miguel Gamborino, a printmaker from Barcelona, experimented with globes and conducted a launch in January 1784. Others, in Arunjuez, Valencia and, later,

Madrid, followed. A French painter named Bouche was picked by the Spanish prince Dom Gabriel to conduct a launch for him on 5 June 1784; unfortunately, just before take-off, a fire broke out, forcing Bouche to abandon the balloon and jump to the ground.[48] A successful manned flight did not take place until the arrival of Lunardi in January 1793.[49] Balloonists also conducted experiments in the Spanish colonies, including some in Oaxaca in 1785 as well as in Vera Cruz.[50] Students of the Royal Academy of San Carlos launched two balloons simultaneously on 17 September 1785 in front of the Viceroy Bernardo de Gálvez and other local officials in Mexico City. The balloons caught fire after rising about sixty feet.[51]

One of the first flights in Russia arose in front of the Winter Palace in Petersburg in November 1783. An early launch in Moscow occurred on 9 February 1784 and, like the Mexico City experiment, ended poorly. In this case, the balloon rose to a height of about 300 feet before falling back down to the ground. A few days later a Frenchman named Demesnil performed another experiment from Moscow during which the balloon travelled over twenty-six miles and, somewhat unusually, landed intact so that it could have been launched again without needing repairs.[52] In German-speaking areas, an early tethered experiment took place on 6 July 1784 from the Prater in Vienna, organized by Johann Georg Stuwer, a Viennese fireworks specialist.[53] Meanwhile, in January 1784 a Benedictine monk named Ulrich Schiegg sent up a hot-air balloon in Ottobeuren. In Augsburg two bookbinders named Bader launched a balloon in February 1784. However, it was not until Blanchard arrived on the scene that people ascended.[54]

Blanchard performed the first manned launches in Hamburg, Leipzig, Berlin and Nuremberg, among other places, but not until 1785 and after. The same is true for the Spanish Netherlands and the United Provinces. A physics professor working at the University of Louvain, Jan Pieter Minckeleers, worked quickly to analyse the properties of hydrogen and, on 21 November 1783, sent up a hydrogen balloon from Heverlé, the home of a local nobleman. On 10 February 1784 he released a balloon from the garden of the Hôtel d'Arenberg in Brussels; this was followed by additional experiments from Anvers and a second launch from Heverlé in May 1784. The Société d'Emulation de Liége organized a subscription in that city at the end of 1783 with other flights taking place in Ostende (16 April 1784), Brussels (two flights in May 1784) and Ypres. Once again, however, it was not until the arrival of Blanchard, at Ghent in 1785 and in Brussels and Liège in 1786, that the first person went aloft in these areas. In Holland an experiment apparently took place as early as December 1783, although little is known about it; another took place on 22 March 1784 in Groningue. Blanchard's flights from The Hague and Rotterdam, however, were the first manned flights.[55] In a letter from 26 March 1784, cited in Tiberius Cavallo's *History of Aerostation*,

Horace-Bénédict de Saussure mentioned using an aerostatic machine to conduct experiments on atmospheric electricity in Switzerland; no mention is made of whether or not an audience was present for the launch.[56]

People in Britain were quick to hear of the invention of ballooning. Zambeccari conducted the first aeronautical experiment; he released a small balloon from the house of Michael Biaggini in Cheapside on 4 November 1783. Biaggini, a maker of artificial flowers, may have helped construct the balloon. Their success encouraged the pair to release a larger balloon on 25 November 1783, this time after first exhibiting it at the Lyceum, for a price, and then charging admission for the event itself. James Dinwiddie followed suit on 18 December 1783 with a small balloon sent forth from the Bowling Green Tavern in London. His success encouraged him to duplicate his experiment from the cities of Bath and Bristol over the next weeks. Aimé Argand, a Swiss savant, conducted a launch for King George III at Windsor Castle on 26 November 1783. This was a very small balloon, only thirty inches in diameter, but it was a great success.[57] By late 1783 and early 1784, similar ascensions were taking place all over England. In addition to Dinwiddie's experiments at Bath and Bristol, experiments occurred at: Manchester in December 1783; Colchester and Derby in January 1784; Oxford Birmingham, Norwich and Sandwich in February 1784; Aberdeen in March; and Macroom, near Cork, in April.[58]

Contemporaries debated over the date of the first manned experiments in Great Britain. L.T.C. Rolt relates a contemporary account that a man named Barber may have ascended on 24 February 1784, but there is little evidence for this.[59] A Scotsman named James Tytler managed to ascend in Edinburgh in what he called a fire balloon, but whether these were flights or, as Tytler referred to them, 'leaps' is unclear. Essentially, Tytler managed to use the balloon to offset his own weight and shoot up into the air for two brief jumps.[60] While Barber's story may be apocryphal and Tytler's efforts questionable, more certain is the success of Vincenzo Lunardi who, after considerable trials and tribulations, launched a balloon in London on 15 September 1784.[61] He had some competition from a medical doctor named John Sheldon, an eccentric who kept his embalmed first wife in his medical laboratory. Sheldon, along with an umbrella maker named Allen Keegan, constructed a balloon eighty feet in diameter in which he made a brief tethered ascent. Sheldon later joined Jean-Pierre Blanchard on one of his voyages. However, it was James Sadler, the son of a confectioner, who became the first Englishman to defy gravity when he soared into the skies on 4 October 1784 from Oxford.[62] Significantly, this date was disputed by some people, such as the early historian of aeronautics Tiberius Cavallo. He noted 'it was found that nobody saw him [Sadler] either ascend or descend'. This serves to emphasize the importance of witnesses for launches, something the Montgolfier brothers knew without needing to be told. In Sadler's case, his first official launch, at least according to Cavallo, occurred on 12 of October 'in the presence of surprising numbers of people of all ranks.'[63]

Richard Crosbie performed the first ascension in Ireland when he released his aeronautic chariot from Dublin in 1785; a French surgeon named Potain also sent up globes in Ireland.[64] The former British colonies in North America, now the Unites States, expressed early interest in ballooning and attempted to duplicate the experiments about which they had heard. Peter Carnes, a lawyer in Baltimore, made just such an attempt in June 1784; however, a gust of wind knocked him from the balloon which subsequently soared to a great height before crashing down in flames.[65] This failure apparently put a damper on American interest in going up in balloons. The first manned launch did not take place until 1793 under the direction of Blanchard.[66]

This survey of the geographic spread of ballooning demonstrates the speed with which this new invention traversed the European continent and abroad during the course of the first year or so after its inception. While France and England have received the bulk of the attention from historians over the last two centuries, it is clear that other areas also conducted experiments. Sometimes these were at the behest of someone with power, influence and money. Thus, both Count Andreani, who worked with the Gerli brothers in Milan, and Dom Gabriel, the heir to the Spanish throne who hired the local French painter Bouche, funded efforts in their areas. In other instances, the impetus for attempts came from local savants, whether amateurs or professionals. A number of experiments took place under the aegis of local professors of natural philosophy. In other cases, such as that of James Sadler, deep interest alone seems to have sufficed to propel someone to give aeronautics a try. While few of these early aeronauts provide information about how, exactly, they came across information about balloons, it seems clear that their sources provided sufficient information for them to copy the technology for themselves. Thus, knowledge about the construction of the frame, the making of the hot air or hydrogen, the materials used to cover the balloon, and the nature of the gondola, all travelled swiftly across nations, continents and oceans with a sufficient level of detail to allow for considerable mimicry.

The pace of this dissemination and appropriation was impressive. The Montgolfier brothers launched their first balloon in June 1783 and experiments in Paris and Versailles followed from August through December of the same year. Fall 1783 also saw similar events in Great Britain and Russia and by the spring of 1784 similar launches had taken place throughout most of Europe. Manned launches quickly followed in a number of places in 1784. Many of these experiments were carried out locally, with the initiative and personnel coming from the immediate area. Those areas with delayed experiments, especially manned launches, generally seemed to have allowed others to come in and conduct the experiments for them. In Lyons, for example, the Montgolfier brothers and Jean-François Pilâtre de Rozier showed up to assist. Blanchard earned his reputation travelling throughout Europe, and North America, and garnered a significant number of honours as the first person to go aloft in a particular area. Thus, the pace of dissemination slowed only if a particular region waited for someone to come from outside the area to perform the experiments.

The speed of dissemination speaks clearly to the ability of people in the late eighteenth century to circulate knowledge and information. The radical nature of ballooning, something hitherto unknown, meant that a successful understanding of the practice of aeronautics required clear, and replicable, information about how to perform experiments. While creating a balloon might seem simple enough – just fill a globe with hot air and release – successful attempts required information on the construction of the globe itself, how to make the frame, cover it and make the cover relatively impermeable. The spread of aeronautics speaks directly to the international nature of the Enlightenment, the willingness to disseminate new ideas, even to social, scientific and political rivals, and the extent to which countries felt disposed to appropriate and experiment with new technologies.

The question remains as to why some areas continued to engage in launches while other areas effectively limited such events or controlled their frequency. Similarly, the incredibly large number of cities in Great Britain and France that conducted their own launches stands out against the rest of the Europe which witnessed fewer launches performed by fewer people. There is no single answer which explains this difference. On the one hand, regional and national differences existed in the funding for launches. Britain and France, and to a lesser extent the United States, tended to pay for experiments through public subscription or ticket sales. This meant that people did not have to wait for the state or for someone with enough money of their own to pay for an experiment; instead, they could pay for it themselves. This sort of argument would suggest that people in Britain, France and the United States were more commercially-minded in general, but also more willing to invest their money in such a transient entertainment as a balloon. By the same token, the Dutch, who certainly can be characterized as commercial, might have been less willing to invest their money in something with minimal financial returns.[67] On the other hand, the continued, albeit waning, interest in ascensions outside these countries can be attributed to the breadth of the audience for balloons in those areas. When the few individuals funding launches in Italy or Spain, for example, stopped doing so, launches dried up; however, even if half or two-thirds of the people in Britain stopped paying to watch a balloon, that still left a large group of people willing to spend their money.

Alternately, the breadth and speed of launches in those areas could speak to the higher rates of literacy and the much larger number of available tools for dissemination, most specifically newspapers, aimed at a large segment of the population but containing scientific information.[68] The public sphere extended further in some countries than others; more people participated, through such devices as the periodical press, and, therefore, more people felt able to engage with the ideas and inventions being circulated. Not only did the British and French, for example, have more money to spend on aeronautics, they were also in a better position to learn about such inventions in the first place because of higher literacy rates and a greater number of periodicals through which such information was relayed.

Deaths and Disasters

The large number of launches taking place across Europe, along with North America, meant, inevitably, that there would be some accidents. Even the experiment conducted by Etienne and Joseph Montgolfier ended with a fire in a vineyard. With aeronauts unable to steer their machines, and with unmanned globes soaring willy-nilly with highly flammable hydrogen gas or small fires on board, problems were bound to occur. The most famous accident was, of course, that of Pilâtre de Rozier, the first person to die as a result of a balloon flight. Along with his status as the first person to ascend in a balloon, this made him the modern-day Icarus and earned him considerable infamy. His story has been told many times. Pilâtre de Rozier wanted to be the first person to cross the English Channel. After Blanchard, along with John Jeffries, beat him to the punch and travelled from Britain to France, Pilâtre de Rozier tried to refocus the attention on himself and become the first man to cross the channel from France to Britain. Unfortunately, the prevailing winds were always against him; to solve this problem, Pilâtre de Rozier and Romains, his passenger, constructed a two-tiered machine. The top globe was filled with hydrogen, which was generally acknowledged to have better lifting power, while the bottom globe was filled with hot air, a substance Pilâtre de Rozier believed gave him better control of the balloon. Conditions finally seemed ideal and an attempt was made on 15 June 1785 before a large crowd of people at Pas-de-Calais. The balloon rose quickly; witnesses then reported that a flame appeared on the top balloon and, an instant later, the entire apparatus was engulfed in fire. Both Pilâtre de Rozier and Romains fell to the ground, having not even made it over the water, and died. Apparently, Pilâtre de Rozier was already dead when people arrived at his point of impact, but Romains had enough life left in him to utter the phrase 'O, Jesu!' before dying.[69]

The cause of the accident is unclear although the risky combination of the two types of balloons is typically identified as the chief culprit. It is possible that a spark from the fire keeping the air hot in the bottom balloon travelled up and ignited the hydrogen in the upper balloon. No matter the cause, the results were devastating for aeronautics. What people had perceived as an enlightened, adventurous, exciting new invention they now viewed as potentially deadly. This helps explain the somewhat reduced number of launches after 1785 as well as the caution with which some states put limits on the ability of aeronauts to perform their experiments.

Not long before his fatal crash, Pilâtre de Rozier had become engaged to a young woman named Susan Dyer. An Englishwoman from Yorkshire, she attended a finishing school near Boulogne, France where they met. Susan Dyer apparently implored Pilâtre de Rozier to give up his ballooning but he would not listen to her on anyone else. Susan attended the launch of his double balloon and watched as he fell to the ground and died. Overcome by the experience she collapsed, dying a short while later.[70] Poor Susan Dyer was not the only one to react so strongly to a balloon accident. An upholsterer's widow named Mrs Saunders, visiting in Islington during one of Lunardi's launches, saw him throw overboard one of his oars, ostensibly used to steer the balloon. She mistook the oar for a body, 'was suddenly taken ill, and in spite of all medical assistance expired'.[71]

The first balloon-related fatality in Great Britain came during an attempted launch by Lunardi in Newcastle-on-Tyne in September 1786.[72] According to his own account, when the balloon was about one-third full Lunardi tried to accelerate the inflation process by pouring a large quantity of 'oil of vitriol' which caused such a quick release of gas, along with a lot of heat and noise. This caused those men helping Lunardi to flee and, in doing so, release their grip on the ropes holding the balloon. Lunardi noted 'several gentlemen, who were holding the balloon, and who were so alarmed, that leaving it at liberty, they ran from the spot.' Lunardi was left alone with the balloon when a young local man, Ralph Heron, grabbed a rope and tried to help. The rope, however, twisted around Heron's hand and he rose to a height of about three hundred feet. 'It is impossible to describe the consternation of the people,' claimed a letter appearing in the *Courier de l'Europe*, 'who saw this young man suspended by his hand from a rope.'[73] Worse, however, the rope broke, dropping Heron to the ground at the feet of his parents and fiancée where he died shortly thereafter.[74] Lunardi wrote that he was 'inconsolable' over Heron's death and was greatly 'affected by the loss of this unfortunate young man'.[75] Lunardi's reputation as a balloonist, at least in Great Britain, never recovered from this accident and he left for the Continent soon afterward.

Although these early deaths were very dramatic, they were part of a large series of accidents, some fatal and others not. Potain went up in a balloon from

Dublin but had some trouble with his landing. A tear in the balloon almost caused Potain to be thrown to the ground. Instead, he became 'entangled in the cords by which the boat was suspended' and 'was dragged upwards of two miles, within a few feet of the earth, and received some considerable bruises'.[76] A Colonel Fitzpatrick ascended in a balloon which burst soon after takeoff; luckily, the accident occurred near the side of a hill and Fitzpatrick leapt from the gondola to safety.[77] James Deeker appeared suddenly over the skies of Bath on 20 April 1785 having taken off from Bristol earlier in the day. The wind played havoc with Deeker's voyage and he found himself 'frequently darted upon the ground'. The balloon suffered considerable damage from the trees and hedges.[78] An anonymous army surgeon in Spain went up in a balloon in Aranjes which 'unfortunately took fire'. The aeronaut fell 'upon the branch of a tree, broke both his legs, and was otherwise much burnt and wounded'.[79] In 1792 a launch led by a Mr Baretu in Aranjuez, Spain ended badly when Baretu 'was seized with a giddiness in his head, fell to the ground, and broke one thigh and two ribs.'[80]

Sadler narrowly escaped serious injury when he was dragged for nearly two miles by a balloon after a 1785 launch from Manchester.[81] Another unlucky aeronaut, Major John Money ascended in July 1785 and quickly blew out to sea. Unable to open the valve and release the hydrogen so he could descend, Money ended up twenty miles off the coast at Yarmouth. His path was seen by people on the ground and they sent boats out to try and retrieve him but without success. Luckily, a ship named the Argus happened upon Money and he was rescued after having spent over five hours in the water.[82] In October 1785 the people of Yarmouth saw yet another balloon pass over their heads, this time carrying the Reverend Peter Routh, Robert Davy and a Mrs Hines, who all fell into the Channel only to be rescued by a Dutch ship.[83]

Zambeccari suffered dual accidents during his flight on 23 August 1804 from Bologna. Immediately after takeoff, a lamp fell over and spilled 'spirits of wine' on his clothes causing him to catch fire but also endangering the balloon and his companion Paul Andreoli. The crowd watching the launch, including Zambeccari's wife and children, 'saw the balloon sail into the sky with its pilot surrounded by flames'.[84] Zambeccari, however, poured water over his head and managed to extinguish the fire. Later, he fell into the Adriatic where he floated for several hours before two fishermen rescued him.[85] In September 1785 Stuart Amos Arnold, with the help of his son and a sailor named George Appleby attempted an ascension from St George's Fields in London. Arnold had Appleby harnessed to a parachute and hoped to treat his subscribers to a launch and parachute drop. The globe, however, did not have enough lifting power and struck some railings; Appleby fell to the ground and the balloon dragged Arnold's son for over a mile until it crashed into the Thames where the boy was rescued.[86] In 1818, Jeanne-Geneviève Garnerin, wife of André-Jacques Garnerin, fell into the Bordeaux River and nearly drowned before being rescued.[87] Francis Barrett consistently had trouble with some of his launches. In one instance the stage holding the balloon crashed before take-off leaving one boy with broken legs and many others severely hurt. In another case Barrett's balloon crashed into a field where it was 'seized by the country people' who had to cut the balloon apart to free him.[88]

Cavallo reported an experiment with a hydrogen balloon, organized by a Mr Gell. After it descended, two men near Cheadle in Staffordshire found it and carried it into a farm house for further examination. Some of the gas leaked out and a candle set it on fire; 'the balloon exploded with a report much louder than a cannon' and 'struck four persons down to the floor'. Although they recovered, they were 'so stunned, as not to be sensible of fire, till they perceived their heads in a blaze: their beards and eye-brows were burned quite off, and their faces terrible scorched'. In addition, 'the windows were forced out with great violence, and the house was otherwise much damaged'.[89] James Tytler, who had nothing but trouble for most of his attempted experiments, suffered an accident during one launch on 29 September 1784 when the wind broke the mast holding the balloon during the inflation process. Two men were on the mast at the time; 'one of them saved himself by clinging to the trees, the other fell headlong, and was taken up for dead' although he later recovered.[90] Some accidents involved animals. In 1817 a balloon was launched from a garden in Paris with a stag in the gondola. The animal fell out and 'was dreadfully bruised'.[91]

In some cases, as with Pilâtre de Rozier, the audience did witness the death of aeronauts. A certain Mr Mosment fell from a balloon in France in 1806. He used a platform under the globe, instead of a gondola, which, as it turns out, was a bad idea. In this instance, as Jacques-André Garnerin reported, Mosment leaned over to let 'an animal drop in a parachute, he lost his balance, and was precipitated to the earth'.[92] In 1802, François Olivari, a physician from Paris, fell from his balloon during a launch in Orléans when his wicker gondola caught fire due to a problem with the furnace he kept on board to maintain the heat of the air in his globe.[93]

In spite of his earlier luck in avoiding serious injuries, Francesco Zambeccari finally died in a balloon accident in September 1812 when a launch from Bologna ended with the globe catching fire and becoming 'entangled in the branches of a high tree.' Zambeccari's companion, a certain Bonaga, survived with some broken limbs but Zambecarri was 'killed on the spot.'[94] At about the same time, a German aeronaut named Bittorf died when his balloon caught fire and crashed into some houses on the outskirts of Mannheim.[95] Windham Sadler, son James Sadler, had a stellar early career and became the first man to cross the Irish Channel from Dublin to Holyhead. However, in 1824 he and his servant James Donnelly found themselves in the grasp of a strong wind which tossed them into the chimney of a house. Sadler was thrown from the gondola and fell to the ground, dying a few hours later.[96]

One of the most dramatic aeronautic deaths came in 1819 when Marie-Madeleine Sophie Blanchard suffered a fatal accident. Marie-Madeleine had taken over the family business after the death of her husband, Jean-Pierre Blanchard, in 1808. Although he had worked full-time as an aeronaut since 1783 and was, arguably, the only professional, long-term balloonist to have survived since the early days, he nonetheless died in a state of penury. Marie-Madeleine decided to pay off his debts and make a name for herself as a balloonist, tasks in which she was highly successful for the next eleven years. She staged regular launches and, when Napoleon Bonaparte needed a replacement for the disgraced Garnerin, she was named the imperial aeronaut. She even survived Napoleon's fall to become the official 'aeronaut of the Restoration'. She had a number of accidents during her career; in December 1811 the *Gentleman's Magazine* actually announced her death during a launch in Rome. Two reports, from 1811 and 1812 suggest Marie-Madeleine passed out due to the great heights achieved by her balloons. One account claimed 'in consequence of the prodigious height the balloon ascended, Madam Blanchard fainted, and continued insensible for some time' while a second report noted that 'after experiencing the extremes of heat and cold ... she fell into a profound sleep, during which her balloon attained an elevation of 12,000 feet'.[97] Her downfall came, however, in trying to please her many fans. She frequently launched globes from the Tivoli Gardens in Paris, a pleasure garden popular during the Restoration. To enhance her spectacles, Marie-Madeleine added fireworks, shooting them off from the gondola. In July 1819 she announced a special spectacle which she then had to delay several times due to poor weather conditions. However, in the end she decided to go ahead with the launch in the less-than-ideal weather. A spark from the fireworks ignited her balloon and she crashed onto the roof of a house, was thrown from the gondola onto the roof and from there onto the street below. She died in front of the house at 15 rue de Provence in Paris on 6 July 1819.[98]

The fascination for accidents was such that some people apparently complained if they did not get to see one. An article in the *Morning Post* in 1785 claimed that 'twelve months ago, the public would hardly believe that a Balloon could possibly ascend without bursting, or some terrible accident befalling an aerial traveler'. Since then, however, accidents were on the wane leaving some people to grumble because Lunardi, at one of his launches, 'did not break his neck!'[99] This is eerily prescient of some modern attitudes toward certain entertainments – such as auto racing – but might also help explain the early and quick popularity of aeronautics. Historians, and contemporaries, often emphasize the spectacular nature of ballooning and the scientific achievement in humanity's ability to conquer gravity and rise up into the heavens. However, the possibility that the people rising up to the skies might come crashing back down could have motivated some of the people attending these experiments. Even during the Enlightenment, attendance at spectacles centred on death, such as executions, drew large crowds and there is no reason not to believe that the lure of a balloon accident might not have enticed some people to watch.

Parachuting

Balloons on their own attracted a large and keen audience, eager to view launches. The possibility of accidents also drew in crowds and, if nothing else, kept aeronautics newsworthy during the first decades after their invention. However, one of the most spectacular feats associated with this invention, parachuting, really amplified the interest in experiments, especially with the advent of the first human parachutists beginning with André-Jacques Garnerin in 1797 (see Figure 1.2). Parachuting was not, of course, a new idea. A number of designs had been proposed over the centuries and a few experiments had taken place off tall buildings. But with ballooning, the concept of parachuting reached its earliest peak and, ideally, added a measure of potential safety to the accident-prone practice of sailing through the skies.

There are a number of early stories of parachute-like events. Jules Duhem cites the example of an infant who fell from a tower in 1591 as well as a steward, Simon de la Forest, who fell from some scaffolding of the church of Notre-Dame-des-Tables in Montpellier in 1650. Both cheated death thanks to their ample cloths and robes which offered an efficacious resistance along the length of a high wall.[100] Leonardo da Vinci conceived of a parachute designed for human use as did Tito Livio Burattini.[101] In 1767 Vincent Lavini escaped from his captivity by the Savoyards using a sort of 'strong umbrella' with which he jumped from the top of a fortress.[102] More modern parachutes appear alongside balloons. Joseph Montgolfier designed a parachute composed of twelve panels sewn together connected to a wicker basket with four inflatable pig's bladders under it. Joseph added the pig's bladders to act as shock absorbers and cushion the landing. He, along with the Marquis de Brantes, used the parachute to throw a sheep over the side of the highest tower of the Pope's Palace in Avignon.[103]

Joseph drew his inspiration for the parachute from the writings of Louis-Sebastien Le Normand although other savants also laid claim to this invention.[104] Blanchard, for example, asserted the idea had been his. In a letter to the *Morning Herald*, written in 1784, Blanchard argued with the claims made by an Englishman named Thomas Martyn who had written about parachutes and argued that he had presented his original ideas to the Prince of Wales in 1783.[105] Blanchard, in response, claimed Martyn's 'Parachute, or Fall-breaking machine struck me not a little I assure you. Had it been an invention of your own, you certainly

Figure 1.2

would have found out an English name for it; but the fact is, that I invented it so early as the year 1777.'[106] Blanchard repeated this claim many times over the coming years although apparently he was the only one to take it seriously.[107] Martyn's response to Blanchard was somewhat vague. He had never claimed to have invented the parachute; instead, he argued that he deserved credit for the idea of applying parachutes to aeronautics although he also noted that anyone who had ever held an umbrella might have recognized the potential for wind resistance.[108]

Although Blanchard overstated his role in the invention of parachutes, he certainly made consistent use of them in his aeronautical experiments. In June 1785, for example, the *Morning Herald*, talking about his journey across the English Channel, exclaimed that Blanchard's 'aerial passage into France was a striking proof of skill and intrepidity' but the article added that it was 'the experiment with the parachute, or large umbrella', that 'was astonishing, nay miraculous to common capacities, and gratifying in the highest degree to the scientific'. Blanchard ascended with a twelve-pound cat, 'placed in a net and connected to the parachute by means of a long cord'. When Blanchard reached a certain height, he 'gradually lowered the parachute from his car' until the parachute opened up. The cat then gradually descended to the ground although the newspaper account did not know where it had landed or if the animal survived the experiment. Nonetheless, the author proclaimed that it was 'incontestably proved, that with a sufficient expanse of silk, or any other stuff, an animal (and here we include human beings) may descend from any height with the greatest security'.[109] Blanchard later offered a reward of two guineas for the return of the cat, but 'the person who found it refused to deliver it up'.[110] In Lille he twice threw a dog overboard, attached to a large chute, the second time before the Prince de Robecq when, in spite of the wind and the rain, the animal landed lightly on the ground.[111] According to the *Gentleman's Magazine*, 'the dog received no hurt'.[112]

Not all of his early experiments went so well. In 1785, while in Great Britain, Blanchard attempted to establish an Aerostatic Academy complete with lectures and demonstrations. He touted an early experiment involving a parachutist who would fiddle while descending. A machine was prepared 'by which he [Blanchard] might have ascended 45 feet; but his heart failed him before he had got to the height of 20 feet'. He jumped off the tower but was not in the air long enough for the parachute to work or for him to play any music. On top of which 'the parachute was broken in the fall, and so was the fiddle'.[113] In spite of this failure with human parachutists, Blanchard continued to work with animals. In 1788 Frederick the Great of Prussia asked to see some experiments. Blanchard used a cat and a bird, which 'were shut up in the basket attached to the parachute', and fastened them to the bottom of the unmanned balloon. Blanchard designed the parachute to separate from the balloon after a time, which it did right on cue garnering a 'very flattering compliment from the King.' The animals landed safely:

> The Cat, who was very unquiet in the state of slavery, appeared to have forgot its voracious nature, having spared the companion of its voyage; and the Bird appeared to have grown bold by the circumstance, and kept perched on the back of its enemy.[114]

The anonymous author who described the event claimed 'in short, the experiment proved the utility of the Parachute'. However, the author must have been either Blanchard or had not been following the debate over their development since he added that the parachute was 'solely the invention of Mr. Blanchard'.[115] In 1793, in Philadelphia, Blanchard conducted another experiment, this time with a dog, a cat and a squirrel, all of whom landed unharmed.[116] After this success, a commentator writing in the *City Gazette* in Charleston, South Carolina, wrote an article suggesting that the precious nature of life had inspired Blanchard 'to invent a machine whereby people exposed to dangerous labors, such as carpenters, tylers [*sic*], &c. might escape the danger of the fall'. The author went on to claim that 'since the discovery of aerostation, more than sixty animals, which he has thrown from the height of the clouds, have descended without the least accident.' Someone with a parachute

> may without the least fear ascend any possible height. A man pursued by enemies or wild beasts, having a parachute in his pocket, might before their eyes, and to their greatest astonishment, evade their pursuit, by throwing himself from the highest rock into the deepest valley.[117]

As the parachutes used by Blanchard were quite large, it is unclear if the author of this article had ever seen one in action. A 1796 newspaper advertisement also claimed that Blanchard had descended in a parachute himself in 1785 in the Netherlands although this was patently false.[118] In addition, not all of his experiments went well for the animals and, in 1797, an unspecified animal died in an experiment in New York.[119]

A number of other aeronauts experimented with parachutes. A British army officer named Colonel Thornton attempted to drop a dog wearing a parachute from his balloon during a 1785 ascension from South Lambeth. Unfortunately, Thornton lacked the necessary skills to succeed and the poor dog 'descended to the earth with great velocity'. The dog survived but broke a number of bones. Blanchard utilized the same balloon immediately after and took a cat with him for his own experiment. He threw the cat out of the balloon and it 'descended gradually, and lighted on a tree.'[120] In 1786, an American newspaper reported on the construction of a balloon in Holland by Mr Don Vanderstruzt. Illness prevented Vanderstruzt from conducting his experiment; his nephew, however, a major named Mr Vac don Bergen, stepped in and ascended in his stead. Bergen, who had helped his uncle in the construction of the globe, took 'with him a parachute, a dog, a cock, and a pigeon, the former of which was let down, and appeared to descend gradually for some time, but afterwards went off in a horizontal course, and became invisible to the spectators'. Bergen also went astray and was lost. The article suggested neither Bergen nor the dog were heard from again.[121] A balloonist named Mr Renault advertised in New Orleans he would launch a balloon on 4 July 1817 'in honor of the day.' Afterwards, he would drop a sheep 'by means of a parachute'; the connection between the sheep and Independence Day in the Unites States remains somewhat vague.[122]

Parachuting really captured the imagination of people, however, after the first human, Garnerin, descended from a balloon.[123] Garnerin had made his first ascent in Metz in 1787 but his aeronautic career was sidetracked during the French Revolution when he fought for the revolutionary army, was captured, and spent two years in prison. After an escape attempt, his rearrest and interrogations, Garnerin was eventually released as part of a prison exchange and resumed his work with balloons. [124] During the Directory and early Napoleonic periods Garnerin earned the title of 'Aérostier des Fêtes Publiques,' or Official Balloonist of Public Festivals, a position he later lost after Napoleon's coronation.[125]

Garnerin had dabbled with parachutes prior to his daring descent in 1797. In 1790, for example, Garnerin sent a small dog aloft in a balloon rigged to explode five minutes after take off. The dog's parachute deployed and it landed twenty minutes later on the rue de Bercy in the Saint-Antoine neighborhood of Paris where it seemed unharmed although it had 'an extraordinary hunger.'[126] Garnerin sold tickets for another parachute drop in July 1790, at a price of twenty-four sous, to take place at 7:00 p.m. on the rue du faubourg du Temple.[127] Garnerin had also conducted a number of launches without parachutes including one in early June 1790. The balloon in this case rose rapidly and to a great height while the crowd shouted 'long live the nation and its representatives' and 'long live the king.'[128] However, Garnerin's real fame came when, on 22 October 1797, he launched himself in a balloon from the Parc Monceau in Paris and, at the height of three thousand feet, separated himself from the balloon in a small basket attached to a parachute (see Figure 1.2). After Garnerin separated the parachute from the globe he floated downwards although, because of the rigid design of the chute, he oscillated significantly. Nonetheless, he landed safely and returned triumphantly to the Parc Monceau.[129]

In the meantime, however, Garnerin earned an international reputation for his parachuting experiments. In 1798 Garnerin conducted another experiment from the Parc Monceau along with a twenty-one year old woman named Mademoiselle Henry, both of whom descended in parachutes near Le Borget. The launch was preceded by several test balloons, an experiment with a globe designed to detonate, and another experiment with a parachuting animal.[130] On the day of the event, Jérome-Joseph Lalande was on hand to help Henry into the gondola. Garnerin reported that Henry's 'firm and assured countenance was admired by the public and inspired him with much confidence.' She even refused the liquors offered to her.[131] After the successful descent, Garnerin claims he could not 'pass in silence that Henry's place was envied greatly by many other young women'.[132] In another experiment, Garnerin sold tickets for a wondrous evening of events including the parachute descent of a 'Vénus aérostatique.' The evening began between 5:00 and 6:00 p.m. with the ascension of a multicoloured balloon designed to help Garnerin determine the direction of the winds and the 'state of the atmosphere.' This was followed by the launch of a hydrogen balloon designed to explode. Then, a large balloon went up carrying a parachute and a 'Vénus aérostatique.' At a particular moment, the globe 'destroyed itself in a magical manner, with flames and explosions' and the Venus would descend in the parachute, just like 'the poets represented the gods leaving Mount Olympus.' Last, Garnerin would ascend, with an unnamed young woman. Tickets for the spectacle cost three francs per person or six francs if you wanted to sit closer to the action.[133]

Garnerin also took his show on the road. He conducted several launches in Great Britain in 1802.[134] On 3 August Garnerin and his wife, Jeanne, ascended from Vauxhall with an Englishman named Mr Glassford. Garnerin dropped a cat wearing a parachute 'which came down from [a] stupendous height' in 'perfect safety'.[135] The following month Garnerin went aloft alone and descended in a parachute before a crowd of five thousand people.[136] Garnerin also performed three flights in St Petersburg and in Moscow, in June, July and September of 1803. Later, in 1804, Jeanne Garnerin made her twelfth ascension in Moscow, along with an unnamed Russian lady.[137] Garnerin continued to make flights throughout the first two decades of the nineteenth century. Garnerin worked closely with his family throughout his career, especially his wife Jeanne and their niece Elisa.[138] Elisa, in particular, earned an international reputation as a balloonist and parachutist, often working on her own. She was just fifteen when she made her first flight. By her twenties she often worked solo. In 1815 she descended from a balloon with a parachute in the Bois de Boulogne in Paris.[139] In 1816 she repeated this feat at the Champ de Mars before an estimated 25,000 people.[140] She also performed in places such as Rouen, Madrid, Padova, and Milan.[141] In 1819 she asked the French government for 15,000 francs to subsidize a launch at a festival for King Louis XVIII, but the Minister of the Interior refused. She was turned down again in 1822.[142]

Other people quickly made parachute drops. During the French Revolution a spectacle in Paris included a parachute. On 20 August 1799 an 'extraordinary festival' was announced for the Hôtel Biron and its garden. The rooms and gardens opened at 2:00 p.m. and at 5:30 p.m. a test balloon would be launched to determine the direction of the wind. Then, at 6:00 p.m. precisely, a physicist, along with a parachute, would climb into a gondola suspended under a balloon which would then be launched, with the assistance of the Turkish ambassador. Once the globe attained a height of around 350 to 400 feet, the unnamed physicist gave a signal, repeated by an artillery barrage, which announced his separation from the balloon at the same moment 'the balloon will ignite over his head and explode'. At this point, the aeronaut would hang 'in the vastness, suspended by his parachute,' and 'descend by a gradually slower movement and return to the garden'. An artillery barrage announced his return. The price for tickets was set at three livres for the garden and six livres for access to the interior. Those wishing access to the workspace paid twenty-four livres for the privilege. There was also an orchestra and fireworks, composed by one the of the Ruggieri brothers. It is unclear if this dangerous spectacle ever took place.[143] In 1799, however, a woman named Labrosse, described in a newspaper account as 'a delicate young female,' went aloft in a balloon from the Tivoli gardens in Paris after which she 'detached herself from the Balloon, and descended by the aid of a Parachute'. She 'fortunately alighted without receiving any material injury.'[144] A Monsieur Margat took off in a balloon from Paris in 1818 and came down in a parachute near the forest of Brevanne; his balloon was later found to have travelled an additional eleven leagues.[145]

****

Balloons captured the imagination of people across Europe. Arguably, however, the general public expressed less interest in the exact mathematical calculations needed for a balloon to rise, or a parachute to descend, than they did in the actual events. This chapter has outlined the ways in which the practice of aeronautics spread across Europe, and some of the reasons why people remained fascinated by launches. The possibility of accidents and the startling innovation of the parachute served to maintain a strong interest in the spectacle of ballooning. This is not meant to suggest that balloons did not also remain popular simply through their innate entertainment (and possibly scientific) value. Not all people attending launches hoped for an accident. In any case, an examination of the geography of ballooning, combined with accidents and parachuting, does serve to illuminate the sustained public fascination with aeronautics. Typically, historians of flight focus on the drop in balloon flights after 1785. In fact, while there certainly was a peak of launches in the first years after their invention, ballooning remains a strong part of the European cultural life through the rest of the century and into the next.

2 THE ENLIGHTENMENT AND THE UTILITY OF BALLOONING

> 'As each one's wit to different actions bend
> Some for utility – some to pleasure tend.'[1]

The invention of ballooning facilitated a grand, if sometimes grandiloquent, outpouring of laudatory poetry and prose proclaiming the 'sublime invention' and the 'philosophical phenomenon' as the greatest wonder of the current age.[2] If the century of Louis XIV had been destined to 'perfect the arts and letters', wrote one anonymous commentator, then it was reserved for the age of Louis XVI to witness 'the grand discoveries in the high sciences.'[3] As Jean Sgard has noted, 'hot-air balloons imposed themselves from their creation as an emblem of the Enlightenment'.[4] Balloons, however, offered the Age of Reason an ambiguous symbol at best. Horace Walpole, for example, lamented that no matter what happened with balloons, posterity would laugh at him and his contemporaries. 'If half a dozen [balloonists] break their necks, and balloonism is exploded, we shall be called fools for having imagined it could be brought to use: if it should be turned to account, we shall be ridiculed for having doubted'.[5] An anonymously published letter addressed to the Marquis de Saint-Just related aeronautics to a number of positive aspects associated with the progress of society: these included the reign of philosophy, the elimination of prejudice and superstition, the abolition of slavery in many of the countries of Europe, the creation of a grand empire in America, and the suppression of the Jesuits by the Pope. But at the same time, other people noted that ballooning still existed in an 'imperfect state' although they frequently expressed the hope that 'the balloon trade [might] lead to some useful and philosophical discoveries'.

John Jeffries summed up this paradox by pointing out that 'some of a sanguine disposition, were led to expect from it [ballooning], very great acquisitions and improvements, in almost every branch of useful science; others, of a contrary temper, after joining in the general applause bestowed on the first discoverer, called in question the utility which it might be of, and whether it would serve any other purpose than that of mere amusement.'[6] For his part, Jeffries claimed that while he did not accept the utility of balloons blindly, he did feel they might at least help illuminate the nature and properties of the atmosphere. In the end, however, the most frequently raised query regarding this new invention was 'what use do balloons serve?'[7]

Some commentators suggested that it was too soon to worry about how people might take advantage of aeronautics. In response to just such a question Benjamin Franklin famously quipped, 'of what use is a new-born baby?'[8] The *Pennsylvania Packet*, citing Franklin, added that perhaps the new child 'may prove a fool; perhaps he may become a prodigy: let us wait and see what his education will produce'.[9] The call to wait and see how aeronautics would develop over time appeared often in the first few years of ballooning. A letter written to the *London Chronicle* by 'A Constant Reader' decried those who too quickly wished to abandon work on balloons suggesting 'it a mark of a little genius, and a narrow heart, hastily to condemn any thing, because we ourselves cannot at once see all the useful purposes to which it may be applied.'[10] Nonetheless, the failure to immediately put balloons to use clearly caused headaches in those who wholeheartedly supported them. Few contemporaries could fail to see the irony in the newest symbol of the Enlightenment being directionless and without obvious use.

This chapter explores the relationship between the Enlightenment, and particularly that aspect of the movement which focused on the pre-eminence of utility and progress, and balloons. Mankind's ability to conquer the skies could not be ignored, even if the people who made it into the heavens found themselves at the mercy of a fickle technology and the vagaries of the prevailing winds. The debate over the usefulness of aeronautics took place between a wide variety of individuals, savants and amateurs alike, across much of Europe and North America. Using newspaper articles and pamphlets, people offered proposals for the potential use of balloons, as well as a number of satires and critiques. This chapter argues that the public debate on aeronautics at the end of the eighteenth century, and the continued problems with finding a solution to the use of balloons, encouraged the continued control of ballooning by individuals more interested in entertaining an audience. On the other hand, savants and those entrepreneurs who wanted to put ballooning to use eventually faded into the background. Both sides wished to find a way to direct balloons, but for the most part entrepreneurs, whose demonstrations and launches did not require the ability to steer balloons, made up the majority of those who continued to pursue the practice of aeronautics. While balloons would find limited use by the military, the inability of anybody to solve the problem of how to steer balloons dominated the debate and ultimately resulted in the failure of this new emblem of Enlightenment.

The Uses of Ballooning

The initial enthusiasm for balloons knew no bounds. In 1784 Fabry, the author of short pamphlet commenting on the flight of Charles and the Robert brothers, claimed that supporters of this new invention treated its critics as ignorant and 'the enemies of useful science, reason, and good sense'.[11] The British coffeehouse crowd even put discussion of politics on hold, or so claimed the *Morning Herald* in early October 1783; only 'the balloon, or aerostatic globe now engrosses the public conversation'. Commentators in France concurred. The authors of the *Correspondance littéraire* claimed that inflammable gas and flying chariots dominated 'all our circles, all our dinners, the toilettes of our pretty women, [and] our academic lycées'. Others readily agreed: ballooning, said one anonymous author, 'fixed the attention of all the savants and became the unique object of conversation in all assemblies', while the *Courier de l'Europe* claimed 'physicists, mechanics [and] all savants' occupied their time with ballooning, something which even the general population recognized as of great importance.[12] If even peasants could perceive the value of ballooning, surely its innate utility would emerge with little effort.

As suggested by the poem that starts this chapter, some people predicated their enthusiasm on the idea that ballooning would prove useful and not just serve to entertain. The notion of usefulness played an important role in the Enlightenment.[13] Lissa Roberts has suggested that the idea of utility in the eighteenth century operated within a complicated matrix of discursive meanings including 'knowledge in the service of the society' as well as its use as a 'watchword for various forms of sociability'.[14] In the case of ballooning, the former definition certainly rose to the forefront of all public discussions. In addition, the notion that the general public could act as witnesses and attest to the scientific nature of the experiment – however loosely defined that notion of 'experiment' might have been – also allowed for an emphasis on the value of group scientific behaviour. Thomas Baldwin, an amateur aeronaut who launched a balloon in 1785 and subsequently published his journal of the event, discussed the usefulness of balloons based on both scientific knowledge and social rank. He wrote:

> It seems a favourite Question, among those who take a Pleasure in objecting to every Thing they neither do nor will understand, to ask, 'Of what Use can these Balloons be made?' and without waiting for an Answer, to say – 'they pick the Pockets of the Public, risque the Lives of the Incautious, encourage Mobbing and Sharpers, and terrify all the World'. These trite Reasonings are all very true, but little to the Purpose: the Effects above described being merely those arising from Novelty. If, says one in an inferior Station; 'they could convert Balloons into common Stage Waggons; Goods might be carried with the greater Expedition:' or, 'into Stage Coaches,' says another: or, 'into Mail Coaches' says Palmer; 'it would be certainly very clever, as I have the Patent:' – 'or into comfortable Carriages to step in out of THE WINDOW, at a

> Moment's Notice; that would be something,' cries a Nobleman: 'it would save one a Couple of Sets of Horses, and would eat Nothing: one might ride one's own Balloon Matches, from one's Window to Newmarket, and from Newmarket to TOWN; dress for Court as we do, and make Nothing of it.
>
> Such are the different Ideas annexed by different ranks of Men, to the Word UTILITY when applied to Balloons.[15]

Baldwin suggested that to focus on the fact that the public for balloons may have been used by aeronauts merely as tools to fund their launches while those same individuals risked their lives overlooked the point. Similarly, both the plebian notion of using balloons for travel and the movement of goods as well as the aristocratic idea of using balloons to replace horses also missed the mark. Ballooning to Baldwin represented an exalted scientific and aesthetic experience, something he tried to relate in his lengthy treatise covering every aspect of his own launch and which included drawings of England as seen from the balloon. In essence, Baldwin did not find it necessary for balloons to have any sort of practical use other than to be balloons. A French lawyer named Chaperon from Guyenne agreed with Baldwin. In a letter written to the *Affiches de Province*, he asked a series of rhetorical questions designed to pull the legs out from underneath the balloon critics. 'What is the use', he asked, 'of waterfalls in the middle of gardens? What is the use of the Opera?'[16] Balloons, he implied, offered their own justification.

Other commentators, less willing to let balloons speak for themselves, often urged patience. Barthélemy Faujas de Saint-Fond, an early supporter of ballooning, believed that further experiments, on a much larger scale, were needed before the true use of aerostatics would emerge. 'But the principle is known', he argued in 1784, 'and it is to be hoped that, sooner or later, it will be proceeded upon, and that some real advantage will be derived from a discovery, the value of which has not yet been generally enough felt'.[17] As early as 1785 one commentator suggested 'the process of ballooning seems at present at a stand, and to rest in a rivalship among shewmen, which of them shall bring [the] most idle people together, and make the most money for themselves'. These showmen exhibited no interest in emulating savants, the author lamented, and did not even try 'to make discoveries.'[18] In 1788 an anonymous letter to the editor of the *Gentleman's Magazine* bemoaned the fact that balloonists seemed to 'have abandoned the thoughts of making further experiments with Air-balloons'. The author suggested that if balloons do not seem useful, it was because they were in the hands of 'ignorant and mercenary adventurers'. If only scientists would pick up the gauntlet and rise to face the challenge of finding a use for balloons, then the problems would be solved.[19] The length of time required before ballooning became useful, some suggested, could be quite lengthy. In 1793 an unknown commentator published some 'Reflections on Balloons' in an American newspaper, in which the author expressed surprise that 'Balloons should be so rashly pronounced, as they are by many at present, mere childish baubles, and at best but objects of curiosity, and incapable of ever answering the purposes of real utility.' People should wait before offering such critiques, the author complained: ballooning was in its infancy and 'it is reasonable to suppose that a century, or less, may mature it into something of solid benefit to civilized man'.[20] As late as 1812, James Sadler rather optimistically pointed out that 'the first labourers in any science can do little more than contribute a few apparently isolated facts'.[21] Thus, a long time might pass before ballooning reached its full potential; those who critiqued aeronautics failed to approach the problem from a true Baconian perspective. Instead, they should welcome the opportunity to continue collecting information about balloons and wait for the day when the new science reached fruition.

While a number of people wrote in general terms about the proposed usefulness, or its lack, of ballooning, many others offered more specific suggestions on how this new invention might be used. As early as 1784 an article in the *Bath Chronicle* commented that 'as to the uses of Balloons, their number stops us'.[22] Another author agreed claiming that balloons 'may prove one of the most novel and serviceable discoveries that this century has produced'.[23] Many individuals speculated, however, on exactly how to utilize balloons. The poet William Cowper even dreamt about the subject: he described a vision where he said 'I drove myself through the upper regions in a Balloon and pair, with the greatest ease and security'. He added that 'the time we may suppose is at hand, and seems to be prognosticated by my dream, when these airy excursions will be universal, when Judges will fly the Circuit, and Bishops their Visitations, and when the Tour of Europe will be perform'd with much greater speed and with equal advantage by all who travel merely for the sake of having it to say, that they have made it'.[24] While Cowper based this list on his dream of flight, many others offered comparable arrays of possible uses for balloons.

In many cases, the lists describing the utility of balloons offered by commentators appeared quite similar. Some focused their attention on a particular area. Piroux, for example, felt that the sciences – including physics, astronomy, geography and mechanics – would benefit most from balloons.[25] The Montgolfier brothers continued to focus their attention on commercial uses. They hoped to establish a permanent route between Paris and the south of France accommodating both freight and passengers.[26] Military purposes were often mentioned. As early as September 1783 the *Morning Herald* suggested an array of purposes to which balloons could be put including using them to transport horses and heavy artillery.[27] The Russian minister in France focused on transportation more generally, whether for military or commercial purposes.[28] Others, like Thomas Jefferson, commented on a wide range of possibilities. In a letter to Philip Turpin,

Jefferson included scientific, commercial and military uses for balloons but also mentioned their use in 'traversing deserts, countries possessed by an enemy, or ravaged by infectious disorders, pathless and inaccessible mountains,' 'the discovery of the pole', 'raising weights, lightening ships over bars', and, 'housebreaking, smuggling, etc.' While he admits that 'some of these objects are ludicrous' he adds that others are 'serious, important and probable.'[29]

Even poets got in on the act with a variety of suggestions. Philip Freneau's poem titled 'The Progress of Balloons', which first appeared in 1784, described how 'The French shall have frigates to traverse the sky' and proposed that balloon transport will allow 'The man who at Boston sets out with the sun, / If the wind should be fair, may be with us at one, / At Gunpowder Ferry drink whiskey at three / And at six be at Edentown, ready for tea'.[30] Even into the nineteenth century these lists had changed very little. In Marco Antonio Costa's 1837 book on aerostatics, he included chapters on the use of balloons for topography, military reconnaissance, astronomy, medicine, natural history, physics, meteorology and public festivals.[31] While some commentators tossed off a few quick remarks about the potential of globes, others felt it incumbent upon themselves to address the problem with greater depth. An early pamphlet, published in 1783, focused entirely on this question with a broad range of emphases, for example, on how astronomers could apply balloons to their field, as well as to military matters. The author claimed the art of war would be the area in which balloons found their greatest use. Most especially, the author hoped, within eighteenth months Gibraltar could be taken for the French from the British.[32]

In general, authors tended to believe that the value of balloons depended on aeronauts' ability to steer them; on the other hand, a few people pointed out uses that did not rely on forcing balloons to travel in a particular direction. Thomas Martyn, for example, suggested 'that one of the obvious purposes to which balloons may be made subservient, is the instant communication of remote events, by night signals'. Watchmen housed in tethered balloons at staggered intervals could light fireworks of various colours, figures and sizes to transmit information across great distances with relative ease.[33] Martyn also believed that tethered balloons could help out the inhabitants of besieged towns as well as armies and naval fleets who want to determine the actions of their approaching foes. Scientifically, 'aerostatic machines, with retaining cords, may likewise be used to throw many new and important lights on the nature, power, and effects of the atmosphere, in its different states'.[34] In addition to such meteorological studies, Martyn suggested that balloons could be applied to astronomical studies.

A few authors, such as Laurent Gaspard Gerard in his *Essai sur l'art du vol aérien*, treated the subject with a startling frankness and honesty. He lists a number of potential advantages balloons may have to offer including airmail, helping cities under siege, enemy reconnaissance, communicating military orders, spreading news from the provinces to the royal court with greater speed, personal transportation and shipping luxury items (such as seafood) from port towns to Paris.[35] Gerard goes on, however, to offer something rarely seen – namely, a list of disadvantages. While a number of critics did, of course, discuss the problems they felt were inherent in aeronautics, few authors offered a side-by-side analysis of the pros and cons. Gerard's list of potential negative aspects related to aeronautics included a wide range of issues. Spies, for example, could penetrate anywhere they wanted into another country. Murderers could use balloons to escape punishment. Thieves could more easily insinuate themselves into, and escape from, the homes they robbed. A dishonest girl could sneak out of the house without her parents' knowledge. On a related note, seducers and ravishers could more easily gain access to their sexual prey. More generally, balloons might cause fires and other accidents; in addition, the boarders of the country would become much more difficult to police.[36] In spite of these serious disadvantages, most of which still required the ability of people to direct balloons, Gerard generally favoured the work performed by aeronauts and had some proposals that would remedy some of the apparent problems. He suggested a sort of flying police or armed force which, in the service of the government, would monitor the skies. He also thought the state should regulate the balloon industry, perhaps with an eye toward controlling who owned balloons. Alternately, balloons could be kept in a public depot and taken out only for specific purposes.[37] Ultimately, Gerard argued for balloons serving a greater good, in spite of obvious problems.

Tiberius Cavallo was perhaps the greatest proponent of ballooning. His 1784 treatise on *The History and Practice of Aerostation* offered an unapologetic glorification of this new technology while severely lambasting those critics who doubted its ultimate efficacy. He wrote:

> Ignorance, curiosity, and often the supercilious wisdom of the splenetic, ask whether it is possible to bring this discovery to be of any use? – and the want of a decisive answer, which it is not in the power of any man to give at present, makes such generally decide against air-balloons; endeavouring to depreciate them still farther by the ridiculous idea of emptiness, which has often been allegorically expressed by the words *aerial, full of air, empty balls*, and *bags full of wind*.[38]

Cavallo goes on to suggest that people have disturbed and checked 'human attention, human life, human labour, human peace and tranquility' in the past and that perhaps this time critics should 'allow some attention to be bestowed upon one of the greatest discoveries of human industry'.[39] Since he was writing his book early in the history of aeronautics, Cavallo did not find it necessary to offer a complete list of the uses of balloons; in addition, he pointed out 'the most obvious uses will easily occur to any person of the least ingenuity'.[40] Small balloons could be used to test the direction of winds, as signals, and to send letters and other small things. Larger balloons could assist people climbing mountains, help individuals who needed to cross a precipice or river, and could be attached to a boat as a sail. The largest balloons, those that could carry more than one person, might be used to ascertain the shape of seas and lands, go to the tops of mountains, convey people from place to place (even if they could not be steered), collect meteorological data, and allow people in hot climates the opportunity to ascend and cool off.[41] 'The philosophical uses to which those machines may be subservient', Cavallo concluded, 'are numerous indeed'. Cavallo hoped that 'the learned, and the encouragers of useful knowledge, may unanimously concur in endeavouring to promote the subject of aerostation, and to render it as useful as possible to mankind'.[42] Later authors frequently borrowed Cavallo's rubric, based on the size of balloons, as well as many of his examples. Jean-Pierre Blanchard and Gardiner Baker's treatise, *The Principles, History, and Use of Air-Balloons*, first published in 1796, appropriated Cavallo's categories for their discussion of the utility of ballooning although they at least cited their source. To these uses Blanchard and Baker also suggested they might 'add the gratification of curiosity and pleasure as a very strong inducement to the practice of an art, in which, with any tolerable degree of caution, there appears not to be the smallest danger.'[43] Blanchard, who spent many years doing balloon 'experiments' for spectators who paid for the privilege of watching, necessarily wanted the entertainment side of ballooning to also appear 'useful'.

Scientific Utility

While Blanchard's 'experiments' seemingly smacked of pure entertainment, others interested in this new technology made a greater show of claiming that balloons would provide an excellent opportunity to expand humanity's knowledge of just about every scientific discipline. As one commentator suggested in a letter to the *Morning Post*, 'hitherto the aerostatic machines have furnished a much greater proportion of amusement and entertainment than of utility'. Nonetheless, the author added, 'every discovery ... carries on the work of Science, increases our acquaintance with the powers of Nature, and enlarges the ability of man to apply those powers to his own advantage'.[44] It was not always easy, however, to see the 'science' in balloons. Sometimes commentators simply assumed that any launch was something in the nature of an experiment and so counted as science. An article in the *New Hampshire Gazette* in 1785, for example, described an ascension conducted by a certain Monsieur Foquett, a resident of Exeter, in which 'large numbers of spectators collected' to observe the 'scientific experiment'. In this case, the event itself was sufficiently scientific to earn the sobriquet of 'experiment'.[45] Similarly, a subscription announced in Philadelphia claimed that the purpose of the launch was the advancement of knowledge toward 'embellishing the principle of aerial navigation'.[46] This understanding of the exercise was not unusual and not unwarranted. As balloonists tested the shapes and sizes of balloons, methods of filling them with air or hydrogen, and techniques for steering them, they were engaging in developing the new science of aerostatics.

When authors talked about the utility of ballooning for science, however, they more typically meant how balloons would benefit other sciences. Thus, these individuals assumed that experiments would take place in the balloon itself, while airborne, and that they therefore required enough space to store and operate scientific equipment along with the ability to use it properly. Savants could perform these experiments in just about any field of science; one amateur suggested that balloon laboratories might soon appear in which 'the experiments of natural philosophy, and of chymistry, which have already been made, may be repeated in all regions, and at all altitudes'. In addition, the author claimed, savants could perform new experiments on a variety of topics including electricity and magnetism.[47] A number of commentators argued for the usefulness of balloons specifically for physics; André-Jacques Garnerin claimed to have made experiments in galvanism during his thirty-fifth aerial voyage in Moscow. In addition, he made an experiment in acoustics by twice firing a musket; he decided that the sound of the gun grew fainter at higher altitudes than at ground level.[48] Similarly, Etienne-Gaspard Robertson, along with a Russian chemist named Sacharoff, made a number of experiments during their launch from St Petersburgh in 1804.[49]

Occasionally, fairly established savants turned their attention to the possible scientific uses of balloons. Joseph-Jérôme Lalande, for example, who expressed sharp criticism of aeronautics in the wake of their first appearance later decided to join forces with Blanchard for an experiment. During the French Revolution, Lalande wanted to analyse the nature of the winds and trace 'their various directions and dangers'. In addition, he believed that 'at the height of 2000 toises ... the stars have no scintillation', a theory which, at that time, could only be proved by using a balloon.[50] Jean-Baptiste Biot, a member of the Institut de France, hoped to perform a number of 'meteorological and physical experiments' from his balloon, rechristened for this purpose as an 'observatory in the air'. Unfortunately, he lost control of his balloon prior to take-off and was unable to conduct his investigations.[51] More successfully, Joseph-Louis Gay-Lussac teamed up with Biot in 1804, launched a balloon from Paris, and took readings of the magnetic field in order to identify any differences in the strength of the field relative to the distance from the earth. Gay-Lussac also made a solo flight a few weeks later in which he redid the magnetic field observations along with additional experiments concerning air pressure and humidity.[52] Some of these efforts at conducting scientific experiments from balloons were frustrated by the crowds of people who flocked to see every launch.[53]

People like Lalande, Biot and Gay-Lussac represent a serious effort by savants to utilize balloons for scientific endeavours; their efforts, however, were hindered by the overall impression given by aeronauts regarding the nature of their experiments. It did not help matters, Cavallo thought, that accounts of ballooning did not limit themselves to recording 'only the improvements' made to the field. Instead, such narratives often related 'the endless repetition of the very same experiment, with perhaps no other change than that of the date, and names of places, experimenters, etc.'[54] More concretely, the lists of the equipment brought along on voyages appear relatively limited in scope and do not always promote the idea that the aeronauts engaged in experimental work. Meteorological experiments tended to dominate the field. John Jeffries, in a voyage with Blanchard, wanted to analyse 'the state and temperature of the atmosphere at different heights from the earth' and 'by observing the varying course of the currents of air, or winds, at certain elevations ... throw some new light on the theory of winds in general'.[55] Pierre Bertholon discussed a variety of experiments regarding meteorology and the instruments savants would need to take with them on their voyages. He thought that individuals might conduct inquiries into the nature of heat with the aid of thermometers and use hygrometers to measure humidity. Barometers, Bertholon imagined, would help balloonists measure the density of the air which would be useful in general as well as to aid aeronauts in understanding how their machines worked in different environments. Eudiometers would help measure the purity of the air. More generally, aerial explorers could take readings on the direction of the winds, test the senses to see how smell functioned at different altitudes, and examine the nature of electric fluid in the atmosphere.[56]

Many aeronauts made a point of including lists of the equipment included in the balloon. Blanchard and John Sheldon staged a launch from Chelsea in 1784 during which Sheldon carried with him a barometer, compass, telescope and flageolet, along with a bottle of wine. However, the additional weight proved too much and the scientific materials were left behind.[57] Jeffries, on a flight with Blanchard, brought with him

> a Thermometer; a Barometer; one of Nairn and Blunt's Electrometers; an Hydrometer; one of Mr. Arnold's Time-pieces; and a Mariner's Compass: Besides which, I took with me a very good Telescope, and several yards of very thin light ribband, coloured, which I could occasionally cut into small bits as I wanted them; and had ready in my pocket a sharp strong knife and scissars.

He also brought vials of water, and some vials (furnished by Cavendish) to fill with air at different altitudes for later testing.[58] Sometimes the equipment included was somewhat vague, as when James Sadler and an unnamed gentleman brought with them 'proper instruments for philosophical experiment.'[59]

In other cases, the scientific nature of the flight was undermined by the other contents of the gondola. Daniel Vanderdrutz brought with him 'a dog attached to a parachute, a cock and a pigeon....two bottles of brandy, a loaf of bread intersected with slices of meat, some slices of ham and sea biscuit, two large fur-cloaks, and a loaded pistol'.[60] Similarly, Vincenzo Lunardi brought 'a cat and a dog, a pigeon in a cage, a bottle of wine, a cold chicken and other eatables'. These accessories make Lunardi's flight sound more like an aerial picnic than an experiment although the presence of the animals could have related to research concerning the effects of altitude on breathing; however, most balloonists who brought animals with them on their flights did so without explicitly offering their rationales.[61] However, in his letter to King George in 1784, Lunardi actually admits that he 'forgot some of the instruments necessary for [his] intended observations' something that 'very much lessened the philosophic importance of [his] voyage'. On that flight, however, Lunardi did remember to bring wine, chicken and bread.[62] Similarly, Richard Crosbie recorded that he flew so high that the 'mercury in the barometer sunk entirely into its globe'. In addition, he found 'his bottle of cordial broke, and [he] could obtain no refreshment'.[63] During Blanchard's voyage over Pennsylvania, the first manned flight in North America, his friends surprised him with a bottle of ether: he noted that he 'took a few drops of it, which refreshed [him] very much'.[64] Thus, with Blanchard flying high over the United States and other aeronauts seemingly more concerned with food and drink than science, it is easy to see how critics could attack ballooning as frivolous.

In some cases people directed their critique at particular individuals, as when Cavallo pointed out that for one of Lunardi's flights he seemed 'to have made no particular philosophical observation, or such as may either tend to improve the subject of aerostation, or to throw light on any operation of nature'.[65] An anonymous author similarly critiqued James Sadler when he suggested that Sadler's voyage was 'of no great use to the public'.[66] Other authors offered blanket critiques. Henry Beaufoy, who hoped to rise above the rest through his own experiments in an 1811 launch with Sadler, wrote that 'few of the aerial voyages that have been performed either in England or on the Continent, have been undertaken with reference to philosophical investigation'.[67]

Military Ballooning

Despite such criticism of their scientific utility, aeronauts continued to make broad claims concerning how balloons might be used. After science, one of the most frequently referenced areas of use was military. The specific use of balloons during the French Revolution will be discussed elsewhere. Here, however, it is important to discuss the ways in which commentators *believed* balloons could be of use for the military. Certainly, everyone thought that balloons could be and were useful in matters of war. As one author noted, commentating on the progress of human knowledge in the eighteenth century, 'as the advancement of science influenced the arts of peace, so it also influenced, in some degree, the miserable art of war. It is sufficient, on this head, just to mention telegraphs, balloons and flying artillery'.[68] This somewhat negative assessment of the place of balloons in aiding military matters echoed an earlier comment made by Horace Walpole who hoped 'these new mechanic meteors will prove only playthings for the learned and the idle, and not be converted into new engines of destruction to the human race, as is so often the case of refinements or discoveries in science'.[69] In spite of Walpole, a number of other thinkers immediately offered positive opinions regarding the military use of balloons. Joseph Montgolfier, for example, thought they might be used to capture Gibraltar for the French while Benjamin Franklin foresaw their use for aerial reconnaissance.[70]

A mere two months after the Montgolfier brothers announced their new invention commentators suggested that globes might have use in times of war. Giroud de Villette, in a letter written to the *Affiches de Province* in October 1783, suggested that the army could use aeronauts to discover the position of their enemy, their manoeuvres, where they marched and how they were deployed.[71] The following month an anonymous contributor to the *Courrier d'Avignon* wrote that 'the utility, the grand utility [of balloons] is for unveiling the route [and] the approach of an enemy army'.[72] John Money, in his *A Short Treatise on the Use of Balloons and Field Observations in Military Operations*, agreed with the assessment that the state could use balloons to examine the enemy's position and movements. He believed there was not a place in England where a balloon or field observer could not be used and also thought this tactic would have greatly aided the British during their conflict with the rebel Americans who frequently managed to hide the size of their forces in the wooded environment in which the battles often took place.[73]

In addition to observing, of course, people in these balloons could also take action against the enemy. Jan Ingenhousz thought that aeronauts could 'throw fire and destruction upon [the enemy's] stores and machines at any time.'[74] In Kingston, Jamaica an unnamed inventor presented a plan to use a balloon to destroy the French navy. The summary of this plan does not mention whether the balloon would be manned, as a sort of eighteenth-century suicide bombing, or unmanned; but the report does say that the plan was rejected as 'England requires no new art to demolish the fleets of her enemies'.[75] One author suggested that people on the ground expressed enough concern about possible balloon attacks to warrant work on countermeasures. In Russia a report circulated about the efforts of one Doctor Germann, living in Lvov, who had invented a set of glasses by 'means of which it is possible to burn aerostatic globes, even though they be 1 and ½ miles from earth in the air'.[76]

In some cases, balloons acquired a stature akin to that of a navy with claims they could function just like their water-based brethren. The *Columbian Centinel* said that balloons might soon be large enough to 'carry 200 men, furnished with 48 pounders'.[77] A British newspaper reported that the Count d'Artois agreed with this report and the ideas of many others, like Joseph Montgolfier, when he argued that balloons 'will prove of great utility against any future attack on Gibraltar; and that it may be classed under the head of a floating battery'.[78] Balloons might certainly be used to send dispatches, carry letters and provisions, and move troops, thus offering considerable support to the standing army.[79] This concern about a floating army continued to worry some people. During the French Revolution a physicist named Thilorier wrote a letter to the government offering to construct a balloon of such vast size that it would enable the army to become portable and which could 'transport [the army] to the heart of England'. While such aerial armies never materialized, the concern was oft repeated in French and British newspapers.[80] In 1799, the *Philadelphia Gazette* satirically claimed that Admiral Nelson had 'converted many French ships into balloons' and no longer worried about any attacks from them except those 'in nubibus'.[81]

Commercial Uses

While scientific and military uses received considerable attention from contemporaries, they perhaps viewed the potential for commerce with the greatest hope and excitement. Although this sometimes included the idea of selling actual balloons or charging spectators for access to launches, most people writing about the possibilities of aeronautics for commerce meant the application of balloons to facilitate the transport and circulation of goods and people. As Joseph Banks suggested in a letter to Benjamin Franklin, balloons could have a greater effect on humanity than anything 'since the invention of Shipping'. Banks saw ballooning as a relief from the constraints imposed by the weight of goods and people and believed 'we shall see it used as a counterpoise to Absolute Gravity, [and] a broad wheeled wagon traveling with 2 only instead of 8 horses; the breed of that Rival animal in some wise diminished, & the human species increased in proportion'.[82] As mentioned above, people thought balloons would make travel swift and easy. Simon-Nicolas-Henri Linguet suggested that commerce with India would be particularly simple although he does not explain why. A director of the English or Dutch East India Companies, he suggested, could 'have breakfast at the Cape of Good Hope, dine and make an expedition to Canton, and return to have supper with his family in London or Amsterdam.'[83]

Immediately after their successful launch in Annonay, the Montgolfier brothers began petitioning the French state for funds to conduct research in the application of ballooning to the transportation of goods. As reported by one commentator in 1786, the Montgolfiers proposed the construction of a balloon that could 'carry sixteen persons' and which would travel between Paris and Marseilles carrying plate glass to the south and reams of paper back to the north.[84] The personal papers of Etienne Montgolfier confirm his deep interest in applying ballooning to commerce and he even succeeded in getting money from the government although he also invested his own money as well. Etienne argued that balloons could solve the problem of localized deprivations through the ease with which balloons could transport necessary goods to those countries and regions in need.[85]

Others also tried to apply ballooning to commerce. A letter published in the *Affiches de Province* early in 1784 suggested that one of the early public launches, from Lyon, marked 'the first age' of 'aerial commerce'.[86] A rumour reported in the *Essex Journal* claimed that in Savannah, Georgia a balloon had taken in '600 bushels of corn' as well as six men on a trip to Jamaica.[87] In Paris during the French Revolution a group of entrepreneurs established a Compagnie Aéronautique. The company launched a subscription hoping to bring in 80,000 livres – ideally selling one hundred sixty tickets at five hundred livres per person – with which they planned to construct an 'aerial frigate'. Although the author of the announcement claimed to have come up with a method for steering the balloon, a crucial element in the success of the venture, nothing seems to have come from this idea.[88] As late as 1817, people still hoped to apply balloons to commerce. Sir George Cayley proposed a public subscription 'for the purpose of ascertaining how far the principle of Balloons supporting heavy burdens in the air may be made useful as a medium of conveyance'. Cayley proposed the creation of a seven-member committee to oversee the project and the 1,000 pounds he hoped to raise.[89]

The idea of fleets of commercial balloons remained a fairly constant trope during the period from 1783 to 1820. In 1800, a North American commentator thought that the French would soon have 'a whole fleet of balloons' which would 'proceed to American upon commercial adventures'.[90] The substitution of balloons for ships seemed plausible to many even after the difficulties in steering balloons became apparent.[91] The *New York Packet* reported in 1785 that a certain Monsieur Vander Monde had proposed the construction of a balloon that could take advantage of the trade winds 'to carry it to any directed point' and that it was possible 'to make the tour of the globe through the air'. If Vander Monde could accomplish this, the author opined, 'he will have discovered the use of balloons'.[92] A man named Charles Diller, writing from The Hague, thought to apply balloons directly to commerce by using them to buoy ships that were deeply laden. This would allow them to enter shallower harbours than had previously been possible.[93] Alternately, an 1818 article seeking to describe life in Boston in the year 2000 claimed that by that time the post office would transport 'all the mails by balloons'.[94]

While some focused on the transportation of goods, others expressed interest in the movement of passengers. In an age where travel, especially overseas, could be long and costly, the idea that travellers might harness balloons to shorten and ease their journeys piqued the interest of many people. In addition to dreaming about balloons, Cowper also wrote a letter to a friend in which he complained of being kept 'a prisoner' due to impassable roads and yearned for the time when 'The Inventors and Improvers of Balloons' would determine a way to use balloons for travel: 'It is always clear over head,' he wrote 'and by and by we shall use no other road.'[95] While Cowper waxed poetical over the possibilities of aerial travel and ignored bad weather, others proposed to implement them. In 1784 a Milanese professor of physics named Venetiani tried to construct a balloon that would 'descend at pleasure, sustain itself with air, or remount without any addition to the machine' with which 'three persons may travel' for periods of two to three days or even as long as several weeks without having to descend for provisions. If these experiments proved positive, the author of the article wrote, then 'it must be allowed that this discovery [aeronautics] is arrived to a great degree of perfection in a short time'.[96] That same year an article appeared in the *Pennsylvania Packet* which suggested travel 'may be conducted with as great expedition and safety through the air, as a vessel on the water, or a stage coach on the land'.[97] The next year the same journal reported an announcement by Signor Zampange and Co. who proposed 'if God permit' the possibility of a 'new real Air Balloon Post Chariot' which could carry four passengers from London to Constantinople in two days with stopovers in Gibraltar, Malta and Lisbon. Each passenger would be allowed fourteen pounds of luggage.[98] While these ideas may have sparked hopes among travellers, in practice the inability of people to steer balloons prevented their implementation. Nonetheless, as late as 1815 some commentators still argued that 'balloons would supersede the necessary use of horses and carriages'.[99]

The Critique of Ballooning

Beyond science, the military and commerce, aeronautics did have other proposed uses. Major McFarland of Haverhill, England, for example, constructed a balloon in which people would be raised and lowered quickly, which had the 'happy effect in putting the blood in rapid circulation, exhilarating the spirits, and promoting health, while it affords a pleasing amusement'.[100] John Mix, working out of New Haven, Connecticut, also suggested balloons had potential medical benefits. He argued doctors 'highly approved' of the motion of balloons and that 'debilitated persons may experience a restoration of health'.[101] Phineas Parker, operating in New York City, concurred stating 'the Properties of this wonderful Construction are as beneficial as they are various affording the most delicate pleasure.'[102] How Mix and Parker arrived at their view that rapidly going up and down in a balloon might have health benefits is unclear as is the possibility that people subjected to this regime might also suffer from motion sickness.

Other potential uses were proposed somewhat tongue-in-cheek or as part an effort to directly critique the very idea of utility applied to aeronautics. Just as the Enlightenment emphasized the importance of the utility of knowledge, its proponents also understood the necessity of questioning received wisdom or of accepting innovations blindly. Thus, unsurprisingly, a fair number of contemporaries felt it incumbent upon themselves to question just how balloons might possibly be useful. One author suggested that this invention would bring about a new change in manners and that young lovers could meet for a tryst by merely putting 'on their air slippers, and, like the deities of old, make use of the clouds by way of curtains!'[103] Criminals might also make use of balloons, some suggested. One satirical author wrote that 'the two foot pads who robbed several carriages on Putney common early on Tuesday morning were seen to go off in an air balloon, over Highgate hill'. The author claimed that the same people 'robbed several persons the same day at noon, near York.'[104] A poem in the *Morning Post* thought smugglers would find balloons particularly beneficial for their trade, 'for in clouds are no custom-house officers found'.[105] A commentator in the *Affiches de Toulouse et du Haut-Languedoc* had trouble deciding how to approach the question. Whether or not balloons were good or bad was too difficult a problem to solve; instead, arguing 'like Doctor Pangloss', the author just settled on accepting that it was at least good that 'all the Globists profited' from the new invention.[106]

The criticism applied to ballooning could take many forms. In some cases, people attacked the behaviour of specific aeronauts; they targeted, for example, the very public and popular launches conducted by Lunardi. The *Morning Herald and Daily Advertiser* talked about the 'Lunardian rage' and accused him of being more interested in ogling his adoring female fans than in balloon experiments. 'The balloon influenza', the newspaper reported, 'is very rife amongst the ladies'.[107] The sexual overtones in Lunardi's launches and general behaviour provided enormous fodder for jokes. In one epigram, 'On some young Ladies wearing Garters inscribed with the name of the aspiring Mr. Lunardi', a poet wrote that 'The pitying tear he had when on high, / And from every fair bosom the heart-heaving sigh, / Now, clasping the thigh of each beautiful Miss, / He has soar'd within sight of the regions of bliss'.[108] Such associations hardly inspired readers to imbue Lunardi with overly exalted, much less useful, goals. That Lunardi brought a cat and dog with him on several of his voyages also led to questions regarding the seriousness of his purpose; one commentator suggested that the dog was as likely to be consulted 'by the learned and philosophic' as was Lunardi.[109] The *Morning Post* added that 'the highest honours are daily paid to the cat and dog of the immortal Lunardi' implying that the animals drew as much if not more attention than the balloon.[110]

While critics attached Lunardi on a personal level, other authors approached their attack through a discussion of risks. Fabry, the author of an early pamphlet of reflections on the new invention, wrote that the 'enthusiasm [for balloons] is unlimited' and that all those who speak against them are 'treated as ignorant, the enemies of useful science, of reason and good sense, etc.'[111] Nonetheless, some people spoke out against aeronautics. Jean-Paul Marat, writing in the wake of the deaths of Jean-François Pilâtre de Rozier and Romains, expressed skepticism about the utility of ballooning and felt that aeronauts acted imprudently just to amuse their audiences. Balloons he felt, were 'pure ostentation' and 'an enterprise without a praiseworthy goal, without utility [and] without importance'.[112] Marat felt that if balloonists exposed themselves to such perils, there should at least be a significant goal to justify their audacity and the risks they took. While threats to life and limb were most commonly mentioned, John Brown lamented over the popularity of balloons in general. He wrote:

> When I read and hear of the modern bustling about air balloons, what multitudes assemble to behold their motion, and from what distance; and think what useful money and still more precious time, is spent in that unprofitable contemplation, it makes me with great grief and shame to think, 'Lord hast thou made men in vain?'[113]

Brown, a minister, described the people who attended such launches as thoughtless and stupid and added they should be more concerned with providing for their families. The Reverend R. S. Medley concurred with Brown's point claiming 'I think an Air Balloon quite fairly can / Be made an emblem of a wicked man!'[114] In the cases of Brown and Medley, balloons were not just useless, but potentially led people into depravity. A letter written by John Newton to the Earl of Dartmouth added to the implicit immorality of aeronautics noting 'we are bad enough already, but were it possible for men to transport themselves at their pleasure through the air, how greatly would the mischiefs and missions of human life be multiplied'. Newton lamented that people admired balloonists for taking a completely useless balloon ride while 'he who came down from Heaven to dwell for a time with men' was 'slighted and disregarded'.[115]

Some critics concentrated their attention on satirizing some of the most common suggestions for utility. Thus, the idea of transporting people and goods received considerable consideration. One wit wrote from London that 'air balloons, if they have any use at all, it must be to translate us to our proper regions'; thus, he added, the Irish will go to 'Taurus, that is, the Bull – the criticks to Scorpio – the ladies to Virgo – the cuckolds to Capricorn – the merchants to Libra, and so forth'.[116] The *Pennsylvania Packet* published an article in which they claimed to be 'happy to hear the balloon has been lately turned to a very useful purpose'. They go on to describe how the publisher of the pro-Irish *Volunteers Journal* was put under arrest but escaped out of a window using 'pocket-balloons'.[117] A poet published some verse on 'The Use of the Air Balloon; or Ways and Means to effect a permanent Union of Parties' in which the most divisive members of Parliament would be tied 'to the aerostatic projection' whose ropes would be cut leaving them to 'blow in whatever direction'.[118]

Similarly, a fake advertisement in the *Virginia Journal* wrote about a new service being offered by Jeremy Fizz-Gigg. He targeted young people 'pining under the controul of maiden aunts, rigid guardians, or fond parents' and had created balloons that would convey people anywhere. The balloons, however, 'will be mostly suited to the accommodation of two persons' although 'if any delicacy should arise' he had extra gondolas to add on 'for the conveyance of chaperons, bridesmaids, bridesmen, or what not'.[119] Other commentators concurred that balloons would enable 'the female rake' who could no longer be locked away behind stone walls as well as 'the wanton coxcomb' who would defy everybody. The libertine 'mounted on his air balloon' takes 'the willing fair one from her window, and flies to what he calls a heaven of bliss'.[120] The implicit licentiousness of ballooning was echoed in anonymous letter written to M. de Saint-Just in Paris. The author wrote:

> Journeys by air will be so speedy, so convenient and so low in cost, that we shall even see women visiting the four corners of the earth, and returning with new fashions in dress. The ladies of the Court will strike up close associations with the Sultans of Asia (and with the Sultanas of Asia) and will take their scented baths and eat ices with these women. The coquettes of Society will take but little time to captivate the hearts of Asiatic and American millionaires. Our gallants of the Court will not be able to resist the temptation to exhibit their graces as far abroad as Pekin and Japan. Finally, our knights, in quest of good fortune, will often deceive the watchful eyes of Eunuchs and even of she who protects the chastity of our Sisters of God in the cloisters.

In addition to all of the frivolousness noted above, the author of this letter also suggested that lovers will be able to escape for trysts and that smugglers will find it easy to sneak goods into walled cities.[121] A poem published in the *Columbian Herald* agreed, claiming that people will soon elope 'to Venus' and that 'all our coachmen' will be ruined.[122] A French print satirized how individuals might use the new gas to travel by air. Instead of using balloons, a 'Bureau des Diligences Aërienne' had employees inject 'gas' into the bottoms of the customers, who then flew away to places as disparate as Peru, Turkey, China and Egypt (see

Figure 2.1

Figure 2.1). Other artists depicted equally silly uses of balloons such as an aerial bullfight or a 'aerostatic grape-picker'.[123]

The Problem of Steering Balloons

> 'Beautiful invention: mounting heavenward, so beautifully, – so unguidably!'[124]

As the quotation from Thomas Carlyle above suggests, aeronauts could have avoided much of the criticism directed against them had they only managed to develop a reliable method of directing balloons. 'Why pursue the expensive, the dangerous experiment?' wrote Anna Seward, when their uselessness had been proved 'by its being found impossible to navigate the machine'.[125] Enlightened thinkers recognized early on that they needed to spend all their energy finding a method to direct balloons or their new symbol of progress and humanity's ability to conquer nature would become simply an oddity that will 'hardly be used, except of a small size for entertainment'.[126] One author, David Bourgeois, suggested the ability to steer balloons 'is absolutely necessary in order to truly enjoy the uses' for which people hoped.[127] Unfortunately, many of the proposals lacked substance. Even in 1783, the Robert brothers wrote about the large number of suggestions made for steering balloons. Pamphlets and books could be found, they said, in every riverside bookstall and café; however, most 'seemed to have been composed to make the ignorance of their authors stand out' and to offer 'the height of ridiculousness'.[128] As late as 1819 Nathaneal Low, in a description of Windham Sadler, Jr's attempt to cross the Irish Channel, could still express hope that a method would soon be discovered. Low wrote:

> It may appear absurd to some persons; but when we consider the infancy of this science, and the extraordinary discoveries made and making with respect to chemical agents, we confess we are not without hopes of seeing, in our own time, such improvements in the art of navigating in the air, as being expressed at this moment might expose us to ridicule.[129]

An early illustration satirized this problem: the image shows an 'infallible method of steering balloons' by attaching the globe to an animal on the ground using ropes (see Figure 2.2). Although ultimately they failed to develop a method to control balloons, it certainly was not from lack of effort; proposals for aeronautic navigation were legion sharing only their complete futility.

Lack of success, however, did not stop people from announcing with great, albeit unjustified, confidence they had succeeded. A report in the *Morning Chronicle* in 1803 claimed 'the art of guiding balloons had lately been discovered at Berne, in Switzerland. Experiments are shortly to be made in England'.[130] No further mention of this discovery appeared. Thomas Jefferson reported with

Moyen infaillible de diriger les Ballons
Air de Pierrot
Air de Pierrot

Figure 2.2

some optimism, in 1785, how 'two artists in the neighborhood of Paris' could 'rise and fall at will, without expending their gas, and to deflect forty-five degrees from the course of the wind'.[131] Unfortunately, it seems likely he was referring to Alban and Vallet, who had helped J. A. C. Charles construct his hydrogen balloons. While they made a number of claims, they never could direct their balloon at will as they hoped.[132] A variety of other balloonists also claimed to have discovered the art of steering balloons. A physician named Dijols la Castagne from Montpellier announced his discovery in 1784 as did a certain Mr Dufour; a clockmaker named Digues put forth his claim in 1786.[133] These people all published their good news, but failed to follow through successfully. A group of men in Madrid attested to the success of the Comte de Galvez in steering his balloon during an experiment in 1784. They described their initial skepticism as well as their surprise at Galvez's achievement which, they felt appears 'to deserve the approbation of natural philosophers, who will doubtless use their endeavours to improve its mechanism as much as possible'.[134] Galvez, however, does not reappear in the history of ballooning and his success was clearly short-lived.

Blanchard frequently claimed to have discovered the art of steering balloons. He often wrote in his accounts of his voyages, as well as to newspapers such as the *Journal de Paris*, arguing just this.[135] Some periodicals lauded Blanchard's attempts while others castigated him for making false claims. In Blanchard's favour, for example, the *Gentleman's Magazine* wrote that he used oars to change his direction and go against the wind while the *Litchfield Montitor* claimed he 'evinced more power over [the balloon] than it was deemed capable of acquiring'.[136] Other commentators took him less seriously. The *Morning Post* suggested that the encomiums lauding Blanchard 'have been circulated from a misconception of his philosophical principles'.[137] In this case, while Blanchard had claimed to steer the balloon he was, instead, being led around from the ground by a long cord. In addition to Blanchard's efforts, Etienne Montgolfier also spent considerable time and money working on the problem of how to steer balloons; he managed to convince enough people of his imminent success that the French government gave him a grant to further his research.[138] Lunardi also made frequent claims to direct his balloons, in his case with oars.[139] Few others agreed with Lunardi that he had successfully navigated through the air.

A number of individuals wrote treatises, some of considerable length, exploring the topic of steering balloons. Some authors wrote in an attempt to claim the 1,200 livres prize offered by the Académie des Sciences, Belles-Lettres & Arts de Lyon for the person best able to 'indicate the best, least expensive, most efficacious and best manner of steering aeronautic machines'.[140] A medical doctor named Salle, for example, submitted an essay on this topic and tried to claim the money. He also wrote to the Marquis de Condorcet, the perpetual secretary of the Académie Royale des Sciences. His system called for a series of flat oars and gears to shift them according to the winds. Both Condorcet and the Académie de Lyon reject his system although they kindly allowed him to revise and resubmit his essay. Instead, Salle publishes his thesis along with the related correspondence.[141]

In addition to those vying for the prize, a number of authors published accounts of ballooning in which they spent considerable time examining the question of navigation. Giacomo Domenico Bruno, for example, proposed a balloon with boat-shaped gondola, sails and oars with which to steer.[142] The physicist Jean-Louis Carra wrote an *Essai sur la nautique aérienne* in which he proposed two three-panelled wings, on which the fabric covering each panel could be shifted depending on the wind. He also included two oars and suggested that the elasticity of the balloon fabric, as well as the nature of the gas filling the globe, be taken into account.[143] Baron Scott's solution to the problem centred on the shape of the balloon. Essentially, Scott argued for a longer, thinner globe.[144] C. G. Kratzenstein, a self-titled 'royal professor of experimental physics' wrote a treatise on aerial navigation in which he summarized various types of balloons, the appropriate dimensions for the globes, and the use of oars with which to steer.[145] An architect and engineer named Carl Friedrich Meerwein published his account of how to fly in the manner of birds in 1784. First, he argued that humans cannot fly like birds because of issues such as their anatomy and gravity. However, he did believe that machines could be constructed, using rotors, which could fly and be steered.[146] The *Monthly Review* provided a brief synopsis of a book by Stephen Calvi in which the author claimed the ability to navigate; their source for the account, however, did not reveal how Calvi purported to do it.[147] Felix Henin published a brief report in which he argued that the force of resistance applied from a reversed parachute could influence the vertical and horizontal direction of a balloon. It remains unclear, however, if he ever tested his theories.[148] When read alongside satirical suggestions for steering balloons, such as the claim that a fish-shaped balloon be 'drawn by four eagles' which the aeronaut has 'trained for the purpose of assisting him in navigating the atmosphere', it is easy to see why some people remained sceptical regarding the overall utility of ballooning.[149]

Bourgeois, in his *Recherches sur l'art de voler*, offered a survey of flight from antiquity to his times as well as an analysis of the various attempts to steer balloons since their inception.[150] Bourgeois categorized these proposed methods into five groups: first, methods in which human force, applied to oars or some other type of simple lever; second, forces 'applied to diverse types of machines'; third, machines that operated on their own volition; fourth, purely physical methods; last, 'physico-mechanical methods'.[151] Despite Bourgeois's view that five categories were needed to describe the various proposals, most of the suggestions can be lumped into one main type. The most common suggestions applied various nautical tools including sails, rudders and oars with the gondola shaped like a ship or the balloon constructed to resemble an animal (most often a fish or bird). The power used to direct these balloons came mostly from the humans on board (in the case of oars) or wind power (for the sails).

Over the first thirty to forty years of ballooning this range of methods changed very little. A pamphlet appeared in 1812, penned by an author named Luzarche, proposing two ways to steer balloons: the first used a mast with lots of sails and a small rotor for direction and speed; the second machine used a prop, oar and parachute.[152] Although the author claimed these were new machines, they differed little from the options that had appeared and reappeared since 1783. Similarly, the *New-Hampshire Patriot* reported in 1816 that a balloon in the shape of a fish 'had been made to move in a horizontal line direct, and on a circle or curved line at discretion, with the same facility as a natural fish in the water'.[153] Animal shapes, including fish but also birds, had been suggested early on in the history of aeronautics although they did little to help direct balloons. A French engineer named A. J. Renaux claimed, in 1789, to have invented three good methods allowing aeronauts to direct their globes; among these he listed 'an artificial bird'.[154] Samuel Hoole even suggested, because of the shape of his machine, that his balloon be christened the 'Flying Fish.' Hoole also proposed that aeronauts eliminate the gondola, putting passengers on the inside of the globe instead.[155] Even in the nineteenth century, however, proposals remained either vague or impractical. The *Albany Gazette* reported a new discovery, soon to be revealed, made by a civil engineer named M. S. J. Pauly, allowing balloons to be 'steered at pleasure'. Another newspaper reported the arrival of a Frenchman who 'has made an important improvement on the balloon which he thinks will be brought to such a state of perfection, that a man can float about in the ethereal space with perfect ease and fly to any quarter with the velocity of an arrow'.[156] In both these cases, the authors made no mention of any specifics and the reports did not resurface. Jakob Degen conducted a series of experiments in 1810 and 1812 during which he applied a series of machines which allowed him to fly 'at pleasure in all directions' as well as to raise and lower the balloon at will. His machines resembled wings.[157]

****

Savants expended considerable time and energy in both making suggestions regarding the possible use of balloons and attempting to solve the problem of directing globes at will. They did much better, it turns out, with the former problem than with the latter. As such, the symbol of Enlightenment paled somewhat over time as it dawned on people that the greatest achievement of humanity's conquest of the natural world – namely, that people had conquered gravity and flew through the skies – was indelibly marred by the complete inability of people to control the course of the balloon once they had launched it.

Consequently, balloons remained in the hands of entrepreneurs rather than savants. More accurately, the individuals who chose to pursue aeronautical studies tended to be those outside of mainstream scientific activity. They received their funding from their audience rather than through traditional patronage networks such as nobles, scientific academies or the state. Had aeronauts found a method of steering balloons, it seems likely that state and industry would have taken a greater interest in their development. Lacking such control meant that balloons devolved into entertainments rather than experiments although the line between these two notions overlapped in a variety of ways.

3 BALLOONISTS AND THEIR AUDIENCE

While historians may approach the subject of aeronautics through the necessary technological changes needed to launch balloons or through the cultural response to this new invention, at heart the history of ballooning is a history of balloonists and their audience. Aeronauts practised this new science and formed relationships with spectators who, whether as customers paying for a new form of entertainment or as witnesses to a scientific experiment, played a crucial role in ascensions. The very size of the audience for some launches, such as the 400,000 people, approximately half the city of Paris, who turned out to watch Charles and Robert ascend from the Jardin des Tuileries in 1783, clearly indicates the social breadth of the people willing to engage in this activity. Even smaller launches often drew crowds of thousands of people, some of whom had paid for the privilege of attending although most simply found suitable spots – rooted in gardens, perched on rooftops, and straddling atop walls – from which to enjoy a free view of the balloon once it had left the earth. Balloonists also came from many social backgrounds and included some savants but also a large number of scientific amateurs. Thus, aeronautics threw together men and women from mixed social and economic circumstances, disparate educational backgrounds, and extremely varied understandings of how this new invention worked and why it might be important.

This chapter explores the social side of ballooning through an examination of the people involved on all sides of aeronautics. First and foremost, this begins with a discussion of the balloonists, their training and background, and their goals in engaging in this practice. These are the individuals who, on the surface, appear to be in charge of ballooning although, as we will see, their control was limited by their inability to finance their projects independently. Dependence on patrons or subscribers took much of the power from them and placed it in the hands of the supporters. Thus, the next part of the chapter examines the audience. Broadly speaking, the audience included the individuals who ascended in balloons alongside aeronauts as well as the vast number of spectators who filled the streets and squares of Europe and North America attempting to get a glimpse of a globe in flight. Although there is no way of knowing exactly who turned out to watch, a number of people commented on the social composition of the audience which gives some indication of who attended these events. In addition, contemporaries frequently described the moment of the launch, and the audience's reaction, which provides insight into the relationship between the balloon and the people who acted as witnesses.

Significantly, women participated a great deal in aeronautics, both as balloonists and as audience members. This will provide some possibilities for exploring the gendered side of scientific practice during this time period. In addition, class distinctions frequently appear in descriptions of ascension, where a wide variety of people turned out, but also at balloon landings. These landings, which often took place unannounced in the countryside among an unprepared peasantry, commonly provoked a somewhat elitist commentary from the city-folk who unabashedly, if perhaps undeservedly, laughed at the reactions of their country cousins. Not all launches succeeded, of course, and riots occasionally resulted from these failed attempts. Therefore, this chapter also explores the social upheaval that arose when the audience for science took matters into their own hands and acted out their own 'balloon launch' complete with ascension and 'crash landing' that might end in the burning and destruction of the balloon.

Balloonists

Compiling lists of balloonists began to occupy the observers of aeronautics as early as October 1784 when, for example, the *Journal de Paris* published a catalogue of the first twenty-four manned ascensions that had taken place since November 1783 when Pilâtre de Rozier and d'Arlandes had ascended from the Château de la Muette.[1] In Britain, the *Gentleman's Magazine* published a list of similar launches borrowed from the *Journal de Paris* but with some additions.[2] Other compilations appeared in the *Scot's Magazine* and the *Monthly Review*.[3] In general, these early lists provided readers with a list of flights along with other pertinent information including: the date; whether the balloon was filled with hot air or hydrogen; who went on the flight; the starting point; the height attained by the balloon; the distance travelled (from 1 ½ miles to over 150 miles); and the length of time the balloon stayed aloft (ranging from seven minutes to over six hours). Although not every entry contained all this information, each did generally include the most basic data including the location and who conducted the event.

From these early lists a few points can be made regarding the first aeronauts. Unsurprisingly, considering that ballooning originated in France, the vast majority of the first balloonists were French. Of the thirty-nine individuals named in the *Gentleman's Magazine* article, thirty-two were French with only four Italians (three from Milan and one originally from Lucca), two Englishmen, and one

American.[4] This corresponds somewhat to the locations of these early flights. Of the twenty-eight successful launches (defined in this case as at least getting the balloon off the ground) nineteen took place in France. Six of these flights occurred in Paris or its environs – namely Versailles and St Cloud – while the other thirteen took place in cities throughout the country including Lyon, Dijon, Marseilles, Strasbourg, Rouen, Nantes and Bordeaux. Interestingly, the first flight in Spain was undertaken by a French painter named Bouche who, at the behest of the infant Dom Gabriel, attempted a launch. The balloon caught fire almost immediately after take-off and Bouche escaped with some difficulty. In addition, the first flight in Rhodes was undertaken by two Frenchmen. The Italian Vincenzo Lunardi successful ascended in the first manned launch in Britain, much to the chagrin of the British who would have preferred a native hero. The other early flights occurred in Milan and the United States of America. All of the balloonists mentioned in this article were men with the exception of Madame Thible, the first women to ascend in an untethered balloon.

Later compilers, such as Monck Mason, Jules-François Depuis-Delcourt and Christopher Hatton Turnor, added more information on these early aeronauts in addition to expanding the lists up through the mid-nineteenth century.[5] They developed a list of fifty-seven aeronauts from the first flight in 1783 through the end of 1784. Of these, forty-one, or about 72 per cent, were French. However, Turnor and Depuis-Delcourt added four balloonists from Vienna as well as some additional British and American aeronauts. It is unclear what criteria were used in creating their lists. They include, for example, James Wilcox as the first American to ascend in a balloon in 1783; Wilcox reportedly ascended in Philadelphia using several smaller balloons. This story, however, was a hoax.[6]

Nonetheless, we can accept much of what Mason, Depuis-Delcourt and Turnor have to tell us about the aeronauts listed. For some individuals, for example, they have identified their status or occupation. Of the people listed in 1783 and 1784 eleven were savants of some kind. The American John Jeffries was a medical doctor while the Englishman John Sheldon was a Professor of Anatomy. Over half of the savants were listed as professors whose specialties including natural philosophy (four), belles-lettres (one), and philosophy (one); one of the professors was also an abbé. The remaining two savants were chemists. Seven of the early aeronauts were members of the nobility of which five held military posts. After these two classifications, the numbers drop somewhat. There were three each of engineers/mechanics, merchants, government officials and architects. Two were painters and one man's occupation was listed as optician.

While the French dominated early ballooning, the numbers change over the course of the first decades. An analysis of the list of aeronauts from 1785 to 1820 reveals that of the nearly 200 individuals listed almost 25 per cent of them were British while about half were French. Other regions included in this expanded compilation include Italy (twelve), Austria (ten), other German-speaking areas (eleven), Russia (four), Poland (three), and the United States, Netherlands, and Denmark with one each. This indicates the beginning of what would be, by the mid-nineteenth century, a significant shift away from the early French control of ballooning. Thus, while the French continued to lead the field, British participation was on the rise. Aeronauts from other countries were also increasing in number. By 1850, the roll call of aeronauts since 1783 included about 500 names of which 313 were British, 104 were French, followed by eighteen Italians, seventeen Germans and a handful of other nationalities.[7]

Unfortunately, for the period after 1784, Turnor, Depuis-Delcourt and Mason provide insufficient information regarding social status to offer any significant discussion. Of the forty-seven individuals whose status or occupation was listed ten were nobles with an additional twelve people having military positions. Six of the aeronauts were savants, including three professors of natural philosophy, and another six individuals were doctors (including one surgeon). There were three merchants and three manufacturers (two of whom manufactured chemicals); there were also two fireworks makers. The remaining occupations included one each for government official, mechanic, clockmaker, carpenter and abbé. This range of occupations and status indicators for the period from 1785 to 1820 matches that of the first year in most respects. The gender of the aeronauts, however, changes more clearly. While only one woman went up in an air balloon in the first eighteen months after their invention, twenty-two women had participated in aeronautics by 1820. Most of these women were passengers in someone else's balloon. For the entire period from 1783 to 1850 Turnor claims forty-nine women took to the skies 'of whom 28 are English, 17 French, 3 Germans, and 1 Italian'.[8]

Four women, however, operated independently or, more frequently, as part of a family group. Marie-Madeleine Sophie Blanchard, for example was the wife of Jean-Pierre Blanchard. After his death she took over the business and rose to the rank of official aeronaut of public festivals during the Directory period of the French Revolution and, later, worked as Napoleon's official aeronaut.[9] Wilhelmine Reichard, a 'berufsballonfahrerin' or female balloonist, undertook seventeen flights starting in 1811 and was the first women to ascend in a balloon in the German states. Her husband was a chemist and aeronaut named Johann Carl Gottfried Reichard and she performed several launches with him; she guided the balloon while he conducted experiments.[10] She also oversaw some of her own launches such as the one in Konigsburg in 1812 when violent winds tossed her balloon around and caused her to crash land. Reichard fainted and was discovered, 'half dead', by some peasants, who had found her on a rock, with the remains of her balloon.[11] Despite this misadventure, Reichard ascended again in Hamburg in 1816 and from Aix-la-Chapelle for the King of Prussia in

1818. A newspaper article described her 'courage and indifference' to the danger she faced and noted she 'entered the basket with the greatest composure and some grace'.[12]

Jeanne Garnerin and Eliza Garnerin, respectively the wife and niece of the aeronaut André-Jacques Garnerin, performed a number of launches with him as well as on their own thus establishing themselves as something of an aeronautical dynasty.[13] Eliza Garnerin was especially active; she engaged in launches in Spain, France, Milan and elsewhere.[14] While not all of her ascensions went well (she was forced to hide after one failed in Madrid), Eliza generally performed to high standards and was occasionally the object of laudatory poems.[15] In addition, she combined ballooning with parachuting such as her descent in the Bois de Boulogne in Paris in 1815.[16] Since her uncle was the first human to descend in a parachute, this was something of a family tradition which she carried out with great aplomb. As one newspaper reported, Eliza, 'the fair aeronaut' jumped out of her balloon and 'then executed a majestic descent near the place from which she had risen, and there received the congratulations of the public'. Afterwards, she was escorted to a play 'by several gentlemen who attended the descent of the parachute'.[17] This behaviour – of risking death and then casually getting out of the burning car, meeting some people and going out for drinks – made Eliza a popular figure in early nineteenth-century Europe.

Historians have gleaned additional information about some of the aeronauts not included in the early compilations. James Sadler, for example, was the son of a pastry chef, something that rather galled the elite in England who would have preferred that the first English aeronaut come from the upper classes. This was a double blow since the first man to ascend in a balloon in England was an Italian, Vincenzo Lunardi, and Sadler's rival for the first British man in a balloon was a Scotsman named James Tytler.[18] Snide comments about Sadler's background occasionally appeared even as people praised him for his efforts. The *Morning Post*, for example, referred to Sadler as 'the little Oxford Pastry Cook' while simultaneously hoping he beat out Blanchard as the first person to cross the English Channel from Britain to France. If he succeeded, the paper opined, his return to Britain 'will greatly reduce the price of *French pies*'.[19] When Blanchard, despite Sadler's best efforts, succeeded in crossing the Channel, the same newspaper reported 'there were great rejoicings at Oxford on account of Mr. Sadler's disappointment, as it is generally believed that it will remove his balloon frenzy, and restore his cheese-cakes and apple puffs to that degree of unrivalled excellence they were once so famed for'. The paper added that Sadler failed in his attempt to launch a balloon across the Channel because 'conjugal tenderness' led him to return to Oxford to 'assist Mrs. Sadler in the labours of Twelfth-day – and thus was English glory sacrificed to Plum-cake!'[20] Nonetheless, Sadler and his two sons, John and Windham, formed a ballooning dynasty that lasted from

1784 until Windham's death in 1824 when he fell from a balloon due, perhaps, to bad weather.[21]

While compilations such as those used above attempted to catalogue every person who ascended in a balloon, inevitably they missed some people. In addition, it remains more difficult, if not impossible, to compile a list of people who launched unmanned balloons. These certainly took place frequently and, since they could be much smaller and required less expertise to manufacture, might be done by a broader range of citizens. However, after the advent of manned launches newspapers and commentators became less likely to comment on any other type of ascension unless it was associated with a particular event, such as a festival, or was a particular form of entertainment. The latter case included Enslen's balloons, shaped to imitate figures from classical mythology, such as Pegasus.[22] Another problem with these compilations is that they fail to tell us how many launches each aeronaut conducted. While some individuals participated in only a few ascensions, and in many cases only one, others performed them with great abandon and a few even attempted to do so professionally. André-Jacques Garnerin, for example, did over fifty launches, while Jean-Pierre Blanchard executed over sixty and Charles Green, in the nineteenth century, performed 249. Other members of the Green family carried out as many as 535 ascensions.[23]

Regardless of their social origins, balloonists were the pop stars of their day.[24] A short impromptu written 'by a Lady' which responded to the query 'what was the strongest proof of a daring mind', by claiming 'Nor was he much inferior who could dare, / In a balloon to sail through liquid air'.[25] Faujas de Saint-Fond described Charles and Robert as 'demi-gods'.[26] People on the ground often tried to follow the path of the balloons through the air and, when they landed, would bring the aeronauts back to the launch site to be feted and honored. Blanchard, after his fifteenth voyage on 3 October 1785 from Frankfurt, was greeted by the local town leaders who arranged for him to use one of the carriages of the local prince for his return to Frankfurt. Once back, he was taken to a theatre where he was received with music and 'general acclamations'. Two 'charming actresses, representing the graces' read couplets to him written just for the occasion, and crowned a bust of Blanchard which was then elevated in the middle of the theatre. Over the next two days Blanchard received yet more attention as well as fifty gold coins, minted especially for the occasion. He, in turn, gave the flag he carried with him on the voyage to the Frankfurt magistrates. The entire process took several days and involved all of the chief citizens of the town.[27]

In addition to being honoured in the period immediately after the completion of the flight, aeronauts were also lauded extensively in print. Poets wrote countless odes in their honour, many of which were reprinted numerous times in different newspapers. After one of Lunardi's voyages in Scotland, for example, Tytler, the would-be Scottish aeronaut, composed some verses:

> Etherial Trav'ller, welcome from the skies!
> Welcome to earth, to feast our longing eyes!
> Once more we, trembling, for thine absence mourn'd;
> Once more we bless thee from high Heav'n return'd.[28]

Tytler goes on to contrast Lunardi's triumph with his own failures and includes a lengthy footnote in which he describes his misadventures with aeronautics.[29] Another poet, Manoel Maria de Barbosa du Bocage, also wrote in honour of Lunardi in which he compares him to other heroes such as Vasco de Gama, Christopher Columbus and Ferdinand Magellan.[30] The British newspaper the *Morning Post* included a number of poems written for Lunardi. In one ode, the anonymous authors claims 'Whatever future merit men shall claim / From guiding the aerial frame, / Lunardi in those annals fair shall stand / The first who led the way from Britain's favour'd land'.[31] In another poem, by the unnamed J.W.C., Lunardi is referred to as 'the dauntless conqu'ror of the realms of air'.[32] More prosaically, but perhaps more common, a flight by Windham Sadler ended with the local inhabitants offering him every kind of assistance they could. Sadler was particularly thankful 'of the hospitable attentions of Captain Skinner, to whose house he was invited, and where he passed the night'.[33]

When Blanchard and John Jeffries arrived in France after crossing the English Channel they were feted over a series of several days. Immediately after landing 'a number of peasants came up, and some horsemen' who placed the balloon in the gondola 'took it up, and proceeded in triumph with it on their shoulders'. When they arrived at the local chateau of vicomte Desandrouin they were greeted at the entrance of the hall 'by a young lady singing some stanzas in honor of [their] enterprise'. The next day the Mayor and Aldermen of Calais along with the King's Procurer General and 'all titled and principal people of the place and neighborhood' congregated to present Blanchard 'with the freedom of the city in a gold box'. Further honours, dinners, meetings with important people, and events continued when Blanchard and Jeffries arrived in Paris and then Versailles.[34] The passengers of the Lyon flight of 1783, including Montgolfier and Pilâtre de Rozier, were carried 'aloft on the shoulders of a surging tide of celebrants'. At a performance of the opera that evening they were 'showered with more wild huzzahs. The singer playing Agamemnon produced a crown of laurels which, characteristically, Montgolfier placed on his wife's head, while Pilâtre (competing in modesty) placed his on Montgolfier's'.[35]

Blanchard also served as the object of numerous laudatory poetical efforts. After his successful crossing of the English Channel Marie-Emile-Guillaume Duchosal published a poem describing him as an 'audacious mortal' who 'in an instant penetrated the brilliant home of the gods'.[36] An anonymous poet lauded Blanchard's forty-fifth ascension, and the first manned flight in the United States, with a poem in which the author describes Blanchard's exalted journey: 'How vast the height, how grand the scene / That your enraptured eye surveys'.[37] A poet, and self-titled 'aerostatic physicist' named Luxero wrote a nine-page poem in which he describes a meeting between Blanchard and the gods including conversations with Venus and Vulcan.[38] The *Journal de Paris* published a considerable number of poems in honour of aerial navigation and the aeronauts. The Montgolfier brothers, of course, received their fair share of this. A poem by Abbé Hollier, for example, referred to them as the 'hardy Montgolfiers'.[39] In 1784, a 'Chanson du ballon de M. de Montgolfier' appeared praising both the Montgolfiers and Pilâtre de Rozier.[40] Of course, the death of a balloonist also brought out the poets. When Windham Sadler died in 1824 Nanny Foulds wrote a few lines 'on the Lamented Death of Mr. Sadler who was thrown from His Balloon'. She wrote 'the Hero rode majestic through the air / From cloud to cloud, not thinking death was near'.[41]

Passengers and other Accoutrements

A distinction should be made between, on the one hand, the individuals who designed, helped construct and controlled launches and, on the other hand, people who were simply passengers. Going back to our original list from the *Gentleman's Magazine*, of the thirty-nine named individuals in the first twenty-four manned ascensions about one-third can be identified as passengers who did not participate in the design, construction or inflation of the balloon and whose presence was not required to operate it. This percentage may be too low as it is not always clear who contributed to the experiments and who did not. Many of the passengers probably helped fund the launches. This was certainly true in Lyons for the launch that took place in January 1784. The bulk of the passengers there were local nobles whose patronage helped fund the experiment but whose inability to decide who should ascend in the balloon doomed the launch to a paltry fifteen minutes in the air since it was overburdened.[42]

Not all passengers were human. A number of aerostatic animals took to the skies. Famously, the first creatures ascend were a rooster, sheep and duck dispatched from Versailles in 1783. On 16 January 1784 Count d'Albion launched a balloon from his garden at Franconville with a wicker basket containing two Guinea pigs and a rabbit; when the globe was recovered, 'the animals were found not only alive, but perfectly well'.[43] Less fortunately, a cat fastened to a basket under the globe launched by Gellard de Chastelais in February 1784 was found dead although 'nobody could guess at the cause of its death'.[44] Other animals made the journey as well. Vincenzo Lunardi frequently took animals with him.

In 1784 he took a cat and dog with him on a flight and they, along with the balloon, were exhibited afterwards at the Pantheon.[45] An American newspaper, the *Pennsylvania Packet*, printed a letter to the editor which claimed that Lunardi left the cat with a woman in Southgate and that ever since 'men, women and children, have beset the old woman's house from morning til night'. The cat was so popular, the paper claimed, 'Balloon Puss is to have a silver collar, with the day on which she mounted, engraved upon it'. The author of the letter wrote a short poem honouring the cat: 'The cat took a sleep, / Whilst the dog had a peep, / And the cat of all cats, / Is Lunardi's sweet cat'.[46] Richard Crosbie's unmanned ascension from Dublin in September 1784 included a cat in the gondola, 'properly secured with netting', presumably so it would not try to escape. This was a tethered launch prior to the main event.[47]

Blanchard also took animals up with him on occasion. For his North American flight, he wrote in his account that he took along a dog:

> A little black dog, which a friend had entrusted to me, seemed to feel sick at this height', Blanchard wrote, and 'he attempted several times to get out of the car; but finding no landing place he took the prudent part of remain quietly beside me: the whining of this little animal raised nevertheless reflections in my mind, which would have affected me very much, had not the view of the country, whose vast extent was expanded before my eyes, opened my mind to softer and more agreeable contemplations.[48]

In a later launch, Blanchard also carried some unspecified animals with him although he does not indicate if they were there for companionship or for some other purpose.[49] Jeffries, accompanying Blanchard, took with him a small dog which huddled at his feet and shook because of the cold.[50] Of course, the animals did not always have an easy time of it. A voyage planned in Ireland in 1785 was to include an aeronaut named Durry, his sister and a dog which 'will be suspended and let down' in a parachute.[51] Blanchard, in 1789, also demonstrated the idea of a parachute with a dog before King Stanislaus Augustus of Poland.[52] Similar experiments took place with cats and a sheep. Animals appeared at other times as well. An epistle, written from the point of view of John Sheldon's lapdog who travelled with him in Blanchard's balloon, appeared in the *Morning Post* in November 1784. In this poem, the dog, named Azor, starts off with some inquiries feigning concern for cats who ascended asking 'How is Nero, and how does the sweet Nicolino? / She was ill when I met you just after your flight / But I hope pretty Puss has recovered her fright'. However, the dog follows this up by noting that he had no idea that he 'was doom'd to ascend in a plaguy plague' so soon after. The poem goes on to poke fun at Lunardi, who was fleeced 'while at the Lyceum', and to denigrate a 'balloono-maniac dunce'.[53]

Audience

While the social and cultural make-up of the balloonists themselves can be identified with some degree of accuracy, attempts to analyse the nature of the audience are infinitely more problematic.[54] Nonetheless, some generalizations about the witnesses to ascension can be attempted through a judicious use of advertisements, information about potential subscribers, contemporary descriptions from eyewitnesses and other associated information. Through these sources, we can discuss with some accuracy the size, social differentiation, reactions and gender of the audiences for launches.

One of the interesting points about the crowds at ascensions centers on their enormously varied behaviours. Attending such a massive scientific spectacle would have been a new experience for most and, as it turns out, people were perhaps a little unsure just how they should behave. In one popular illustration, addressed 'to the amateurs of physics', people are clambering over a wall to try and see a balloon take off while some men take the opportunity, using telescopes, to stare at the backside of a woman whose dress has risen up in the back to reveal her globe-shaped bottom (see Figure 3.1). At a time when spectacles in general were undergoing significant transformations in audience/spectacle interactions, and in an age when public performance of political spectacles (such as coronations or civic fireworks) were common, people had a variety of potential approaches they might adapt to launches.[55] Simon Schama argues that popular scientific events like these altered the nature of civic festivals since 'as a spectacle it was unpredictable; its crowds were incoherent, spontaneous and viscerally roused. Yet they were neither a mob (un attroupement) nor a random aggregate'. He goes on to suggest that the audience had 'a kind of temporary fellowship in the open air'.[56] This characterization, especially regarding spontaneity, is debatable since most of the crowds witnessing the moment of lift off were there by design, either as paying customers or in response to announcements indicating the day, time and place of the launch. On the other hand, the idea of the audience experiencing a 'temporary fellowship' has more validity, especially since the audience was often treated as witnesses to an important scientific experiment in the guise of a spectacle.

The overall size of the audience varied dramatically, mostly because of the very large numbers of people who turned out for the early launches. Thus, the first unmanned launch of a hydrogen balloon in Paris, from the Champ de Mars, drew 'an immense crowd'. One author claimed that 'both banks of the Seine were covered by a multitude of spectators'.[57] The first launch of a manned hydrogen balloon in Paris, from the Jardin de Tuileries, drew half the city (about 400,000 people).[58] Zambeccari's ascension from a garden in Norwich in 1785 was conducted in the presence of 'more than 40,000 spectators'.[59] Even in the nineteenth century the crowds could be quite large. When James Sadler did a flight in Bristol the report indicated that the spectators were 'innumerable'.[60] People utilized every possible location for viewing the globes. For one flight the towers of the cathedral of Notre-Dame were 'covered with observers and the curious'.[61] Sometimes the crowds were too large for people to attempt estimating their size. The unmanned flight from Versailles in 1783 took place 'in the middle of a prodigious convergence of spectators'.[62] At 'ten a.m., the road from Paris to Versailles was covered with carriages'. The audience watched from every possible location; 'the middle of the streets, the courtyards of the chateau, the windows and even the roofs, were filled with spectators'.[63] Similarly, the 'immense mob' attendant at the Champ de Mars in 1783 spilled out from the field itself and people also watched from 'the edges of the river, the road to Versailles, and the amphitheater in Passy'.[64] Large crowds also appeared in England. For James Sadler's second launch from Oxford in 1784 'an immense crowd of all ranks thronged

Figure 3.1

the streets, the buildings, the towers, the trees and the fields adjacent to the place of ascent'.[65] Sadler's launch in Bristol drew 200,000 people who swarmed the hills as far away as Bath and who all shouted as 'the balloon rose with the most majestic grandeur'.[66]

Other commentators made clear the breadth of the reception balloons received. In 1785, the *Massachusetts Centinel* claimed experiments with balloons were seized with avidity by 'almost all ranks and denominations'.[67] The qualifier here, 'almost', is probably meant to exclude peasants from the list; unfortunately, however, the author does not clarify which religious groups were disinterested or against aeronautics. A launch planned for Philadelphia in 1784 led the organizers to call for advice from their potential audience. All persons, they said, 'whether philosophers, mechanics, literary or private gentlemen, who are capable of giving their advice and assistance' were assured that their opinion would received 'every proper attention'.[68] The intended audience is somewhat limited although the inclusion of mechanics seems to acknowledge that some practical experience in constructing things might be helpful. Another description also indicates clearly the range of interest in this phenomenon. An anonymous report from London which appeared in Boston's *Independent Ledger* suggested:

> [I]t is fashionable to speak of balloons – my Lord speaks of balloons – my Lady speaks of balloons – Tom the footman, and Betty the cook, speak of balloons – yea, and balloons shall be spoken of. The sprightly Miss talks of nothing but inflammable air, and mamma checks her presumption, and tells her that air balloons and inoculations are equally impious.[69]

Thus, all the members of an aristocratic house, from the nobles to their servants, talked about balloons. Even the 'sprightly Miss' talked about them although her mother suggested they were an inappropriate part of polite conversation. The 'mamma' seems to be the only one not interested in this subject.

Further insight into the behaviour of the audiences can be gleaned from Anna Francesca Cradock's journal recounting her time in France from 1783–6. Cradock attended several launches in the French capital during her stay including the one staged by Charles and Robert from the Jardin de Tuileries in December 1783 when half the population of Paris turned out to watch. For a launch in March by Blanchard from the Champ de Mars Cradock sent her maidservant ahead to reserve a space for herself and Lady Sussex (she also allowed the maid to watch along with someone named Pierre). In this instance, Cradock claims, 'the entire population, some on foot and the others in carriages' showed up to 'assist in the launching of the balloon'. While certainly an exaggeration, the description here indicates the class breadth of the audience. With her maid watching as well as her aristocratic friend, with people on foot and in carriages, we can get a sense of a broad spectrum of people.[70]

Another spectator, a Prince Bariatinsky who was visiting Paris from Russia, noted a broad range of reactions. 'Joy, fear, terror, and enthusiasm', he wrote, 'was portrayed on every face'. When Charles returned to Paris he was greeting with a 'thunder of applause and enthusiastic cries of "Long live Charles"' while 'women threw bouquets at his feet and musicians played fanfares and banged on drums'.[71] Louis Sébastien Mercier commented on this same launch. The members of the 'undulating and [socially] varied' crowd raised their hands to the sky in attitudes of 'surprise, admiration, joy and astonishment'. 'Some of them cried out of fear for the bold physicists', Mercier added; 'others fell to their knees, suffocating in surprise, terror and emotion'.[72]

It is also the case that a kind of sentimentality, or at least an emphasis on the sense experiences, loomed large at launches.[73] Audience members certainly expressed a wide variety of emotional responses to the moment of ascension; many of these responses were far from analytic, unemotional or logical. Thus, although they were technically witnessing a scientific phenomenon, the audience felt empowered to treat it as they would a theatrical spectacle. This included cheering, fainting, silence and other kinds of responses. Thus, when Andreani launched his balloon in Milan in 1784, he did it before 'an immense mob of spectators' who evinced a 'mute silence'. This moment was 'very difficult for the spectators;' they had 'pale faces' and there was a 'universal trembling seen in the anguish and fear from all the sensible hearts' watching.[74] Even after the balloon had taken off, the 'eyes of the spectators, fixed on the face of the new Daedalus, followed him with dread and anxiety'. However, once the globe was in the air, and Andreani seemed fine, the scene changed. Suddenly, the multitude of the citizens broke out in 'applause'.[75] When the Robert brothers conducted a launch from St Cloud in 1784 their efforts were also greeted with 'universal applause'.[76]

While applause was typical, so too were other emotional responses. Pilâtre de Rozier claimed that the men present at a 1784 in Lyon were divided 'between fear and admiration' while some of the women fainted and others fought to hold back their tears.[77] This was an emotional moment for the spectators who felt a 'mixture of joy and fear' as they cried out and clapped their hands.[78] D'Arlandes, in describing his flight with Pilâtre de Rozier in November 1783, wrote that the crowd coalesced as a 'single mass' and 'by a involuntary movement' focused on the balloon as if it were the only object to be seen.[79] The satisfaction felt by the ladies viewing the first launch from the Champ de Mars was so great they ignored the strong and abundant rain that was falling at the time in order to watch.[80] Even the people who looked on as Charles moved the balloon from his workshop to the Champ de Mars felt overwhelmed. The cab drivers stopped their carriages, 'prostrated themselves humbly, hats off, during the time the procession was in front of them'.[81] The *Courrier d'Avignon* reported that all the spectators at Charles' launch felt 'pangs of anguish' and that 'some Ladies were taken ill'.[82]

At Blanchard's launch in Boston in 1795 a similarly loud reception greeted part of the launch. This was an unmanned launch although the balloon did have 'a couple of animals in it' who, when they separated from the balloon, elicited a response from the crowd: 'the spectators evidenced their satisfaction by repeated huzzas'.[83] Sometimes the noise was reinforced by music played before and during the launch. In Nantes, a 1784 flight was accompanied by 'a warlike music' to 'amuse the brilliant assembly'.[84]

While noise was common, it was also typical for the audience to express some sort of astonishment, surprise and awe. A launch in 1790, for example, 'excited the admiration and surprize of the beholders'.[85] Although the commentator does not specifically state that the crowd was silent, or vocal, their surprise would seem to indicate at least that they were taken aback by the sight of the balloon in flight. Similarly, a flight in 1784 caused men and women 'great and humble' to fall 'to their knees'. The flight launched from Lyon in 1784 led the audience to 'hold up their arms and hands by an involuntary movement as if to support the balloon in its fall'.[86] Thomas Baldwin wrote 'he ascended, amidst Acclamations mixed with Tears of Delight and Apprehension, the Misgivings of Humanity, and other usual Sensations of Surprize'.[87] Baldwin also provided some emotion of his own. When he finally was able to kneel and look over the edge of the gondola during his flight he was moved by the scene of 'grandeur and beauty' so much so that 'a tear of pure delight flashed in his eye'.[88]

In some cases the reaction to a balloon incorporated a religious sentiment. This could be an acknowledgement of the god-like qualities of aeronauts or the belief that the devil must somehow be involved. Two Irishmen witnessed a globe in flight and 'their amazement was very great; one thought it was the devil appearing in the clouds, taking the tube for the infernal's tail; the other supposed it an angel coming to warn wicked sinners of the day of judgment being at hand'. After the balloon landed one of the people present approached it holding a torch; unfortunately, the balloon was filled with hydrogen and a spark set it off. This did not help the opinion of those present that the devil was involved. 'One man was burned in a shocking manner in the face, and a woman slightly; some fainted, and others ran away to give notice to their friends of the approach of the diabolical spirit'.[89] A commentator ridiculed the Spanish response to a launch in 1784 noting 'balloons are not much in estimation here' and claiming the spectators 'cried out unanimously that it was the Devil'. A century earlier, the author added, 'this unlucky experimental philosophy would have fallen a sacrifice to popular superstition'.[90] In describing his first aerial voyage in England, Lunardi claimed the items he threw out of his balloon were treated like 'relicks of the most celebrated saint'. Lunardi's first ascension was of such interest to people that a jury quickly acquitted a prisoner and ran outside to see the launch. Afterwards, one of the Judges told Lunardi he had 'certainly saved the life of a young man who might possibly be reformed' after his acquittal.[91]

In other instances, rather than making a lot of noise, the audience might stay remarkably silent. Thus, when Blanchard ascended in Basel in 1788 'a mournful silence, expressive of their inward anxiety seized all the numerous spectators'.[92] Whether they were anxious for the fate of Blanchard or suffered some inner angst about engaging so directly with the forces of nature remains unclear. During the French Revolution the government included flight as part of the ceremonies surrounding the completion of the new constitution in 1791. Ascending from the Champs-Elysées, the globe was part of a much larger entertainment including music and visits from the King and Queen. However, in the midst of the festivities the crowd behaved very calmly toward the launch; one observer claimed 'the beauty of the flight, the quiet, the decency, which never ceased to prevail in the midst of a prodigious body of people, gave to this feast a reputable appearance'.[93] In the midst of revolutionary fervour, a calm ascension seems out of place; nonetheless, this voyage was cause for a sedate response rather than shouting and excitement.

Earlier, in 1786, the author of an article in *L'Année littéraire* commented on his experience at a launch: with his 'eye fixed' he 'observed without saying a word' and he 'could hardly believe his eyes'. 'I stayed a long time in that pose', he added.[94] James Sadler also made note of an instance of silence. In describing his ascension from Dublin during his attempt to cross the Irish Channel, Sadler wrote his balloon moved slowly and was perceived as 'grand', having an 'irresistible effect on the beholders' who restrained their 'burst of approbation'. In other words, the crowd was so astonished by the slowly rising, majestic balloon that they forgot to applaud.[95] The launch led by the Robert brothers with the Duc de Chartres caused those closest to the balloon to kneel on the ground 'at once to breathe an ejaculation for their safety, and to give the more distant [a] better opportunity of enjoying' the scene.[96]

Occasionally, the audience for balloons might be described as maniacal or crazed. One newspaper reported 'the Philadelphians are at present Balloon-mad'. This madness, according to the anonymous author, arose for two reasons. First, the Philadelphia newspapers were replete with notices to and from Blanchard who was about to perform a flight from that city. Second, the madness stemmed from the amount of money spent for a spectacle that, in the opinion of the author, was not particularly useful. Only the novelty of the event can 'justify the sacrifice'.[97] Alternately, the crowd that watched as Blanchard and Jeffries descended in France after crossing the English Channel also exhibited a kind of madness. In this case Jeffries noted that the people were 'mad' to eat the remains of his lunch, simply because it had flown above the clouds. Jeffries commented that the people helping him saw 'in the bottom of our Aerial Car a few bits of chicken, and morsels of bread' and so he divided it among them, at their 'urgent requests' since everyone was 'eager to get some of that food, which they had seen literally descend from the clouds'.[98]

Gender

'I have got a balloon wife. Damn that Lunardi'.[99]

Simon Macsarcasm wrote the line quoted above in a letter to the *New York Packet and Advertiser* in 1785 adding 'She was the first in our parish that was affected with the *balloon mania*: and what a life have I had since! Balloon hats, balloon trimmings, in short balloon everything, even to plums and potatoes'. The letter writer goes on to write about a fight between his wife and the local curate over the advantage of balloons, during which the curate tells Macsarcasm 'Your wife is a balloon'. Musing on this, Macsarcasm concludes the curate is a 'balloon parson' adding such clergymen 'may now be said literally to shew us the way to heaven'. He finishes his letter with a plea to the editors: 'For God's sake, mention [Lunardi] no more in your papers, otherwise, I am afraid, we shall have *balloon bedlams* in every part of the country'.[100] Women were involved in ballooning at every stage, as aeronauts, passengers and members of the audience. However, while this was sometimes a laudatory fact, at other times people lamented female interest and participation in aeronautics. A commentator in London learned a lady intended to ride in a balloon released from Hanover Square in 1784 and remarked 'if she goes, there will be little need of inflammable matter', a comment intended to play on the idea of women talking too much.[101] This joke was repeated in the *Morning Post* which noted that many 'Belles Esprits have intimated to Mr. Lunardi, an ardent wish to visit the higher regions, and have promised to supply a competent quantity of *inflammable matter*' adding that Lunardi 'politely declined' their offer.[102]

Women were certainly frequent passengers in balloons. The first to mount in a tethered balloon were four noblewomen, the marquise de Montalembert, the comtesse de Montalembert, the comtesse de Podenas and Mademoiselle de Lagarde who arose from the Faubourg St Antoine in Paris on 20 May 1784. These women were accompanied by the marquis de Montalembert and monsieur Artaud de Bellevue.[103] The first woman to ascend in a balloon, Madame Thible, did not do so until 4 June 1784, eight months after the first trip by men. Thible, apparently 'singing like a bird', accompanied Monsieur Fleurant in a launch from Lyons attended by the King of Sweden. The balloon was named *Le Gustave* in his honour.[104] Some female passengers were quite young. When Baldwin landed at the end of his long balloon adventure, he asked the people in the area if they wanted to go up in the balloon while tethered. Many 'answered in the affirmative' including 'a fine blooming Girl' who 'mounted her Car with great

Spirit'. Baldwin noted that another person who took advantage of his offer was a 'smart young fellow' who got in and 'was no sooner raised a few Yards above his Companions, then the florid Colour forsook his Cheeks; he trembled; bent himself double with Fright; and the Balloon was obliged to be hauled down'. Thus, Baldwin emphasized the courage of the young girl 'comparing it with that of the Person who had none'.[105]

The first woman to ascend in a balloon in Britain was also quite young. Mademoiselle Simonet, a French girl aged fourteen, accompanied Blanchard in his London expedition of 3 May 1785. Her younger sister also participated in a launch later that same month.[106] Other young women followed, perhaps because they were smaller, and therefore lighter, thus not requiring a particularly larger balloon. An aeronaut named James Deeker was twice supposed to go aloft with a fourteen-year-old-girl named Miss Waller in June 1785. The first time, on 1 June 1785, she could not go because of wind damage to the balloon which could not lift two people. She 'lamented in the strongest terms her not being able to go up with him, and was with great reluctance dragged from the car'. A second flight was planned for 22 June 1785 but she missed that one too, this time due to illness.[107] In France young women were also frequent companions for balloonists. During the French Revolution a young Parisian girl ascended in a balloon with an unnamed aeronaut on 13 July 1790, perhaps as part of the celebrations surrounding the first anniversary of the fall of the Bastille.[108] Blanchard also conducted two flights with the ten-year-old Mademoiselle Maison.[109] Garnerin also ascended with young women, including his daughter but also a 'young citizen' not related to him. This launch, from the Parc de Mousseaux, caused some concern as it was thought improper for 'two persons of the opposite sex to be elevated together in the air'. However, Garnerin was granted permission to do the experiment.[110] In 1816 another young women, aged fourteen, accompanied the aeronaut Bugustin in a launch during the festival of St Louis in Paris.[111]

While younger women often ascended in balloons, more mature women also went aloft. Mrs Sage was the first Englishwoman to go up in a balloon when she ascended with Mr Biggins on 29 June 1785. Her trip was no easy feat since at over 200 pounds she weighed more than the average person at the time and, as a result, more lifting power was needed. Initially, the voyage was to be made by Lunardi and a Colonel Hastings, along with Biggins and Sage. When it became clear the balloon's ability to rise was being overtaxed Lunardi and Hastings got out and left Biggins and Sage to enjoy the flight. They landed an hour later in a field only to be accosted by the farmer who accused them of damaging his crops.[112] A local landlord, Mr Holden, perhaps aware of the problems which faced Biggins and Mrs Sage when they landed, 'promised the farmers in his parish, a remission of half a year's tythes, in consideration of the damage which they might sustain, by so great a concourse of people on their lands'.[113] The theme of women weighing too much for the lifting power of the balloon continued with a launch in Suffolk conducted by the local pastor, Reverend Peter Routh, who proposed to ascend with Miss Fanny Shouldham and Robert Davy. However, when the ropes were untied the balloon did not move. Thus, 'it was judged expedient to lessen the weight contained in the gallery, by Miss Shouldham getting out, and Mrs Hines, who was desirous to accompany the Gentlemen, and less corpulent, taking her place' after which the balloon instantly rose. Additionally, a Miss Grice sought to accompany Zambeccari and Sir Edward Vernon on a flight in March 1785; however, even though they threw out a great deal of ballast, 'after making three or four attempts, the heroine was obliged to give up the pleasure of the ascension'. After she gave up her seat, the balloon soared up 'with amazing velocity'.[114] Apparently, Zambeccari and Vernon lacked the gallantry of Lunardi and Hastings. In addition to Lunardi's chivalric behaviour towards Mrs Sage, he also allowed a Mrs Chisholm to take possession of his balloon after he landed near her home at the end of one of his Scottish flights. Mrs Chisholm wished 'for an opportunity to attempt an experiment in the unknown regions' and so got into the gondola and 'sailed triumphantly for about three miles'. Lunardi followed her on a horse and 'everything was conducted with the greatest decorum'.[115] This last mention of decorum echoes the French concerns about Garnerin and his young female passenger.

Blanchard frequently ascended with female passengers. Madame Tummermans of Brussels accompanied him in Metz in June 1788 and in Warsaw in May 1789.[116] The *London Chronicle* reported that Blanchard ascended in a tethered balloon with several women including Madame Dudonet and the marquise de Brossond. In all, Blanchard allowed nine women to go aloft in his balloon; according to the newspaper, Blanchard reported 'they behaved with so much spirit that they would have been excellent companions in difficult voyages'.[117] Garnerin also had several female companions on his voyages. In addition to the unnamed girl mentioned above, a Mademoiselle Henry went with him for a 1798 flight. Garnerin reported that his 'beautiful and brave companion' was not alarmed when the balloon began to be agitated after ascending above the clouds.[118]

Several women tried to ascend in balloons but could not. One unnamed women offered the Robert brothers fifty louis for permission to make a voyage with them, but she 'was politely refused' with no explanation given as to the reason.[119] Thirteen-year-old Miss Port wrote a letter to her father in 1785 in which she describes going to see Lunardi's balloon. 'I should like to go up with him', she wrote, 'provided he does not cross the sea'.[120] Unfortunately, it remains a mystery as to whether or not her wish was fulfilled. Similarly, the fate of 'an English lady of spirit, beauty and family' named Litchfield who 'expressed a desire to take a tour in a balloon' remains uncertain.[121]

As noted above, balloonists sometimes had to defend the morality of their launches when women were present. This concern is mirrored in the rampant sexuality associated with the practice of aeronautics. Contemporaries had already noted that women responded viscerally to witnessing a launch, sometimes fainting or crying out; how much more would they be affected by taking part in a flight themselves? Blanchard's companion, the young Mademoiselle Simonet who was 'said to belong to the opera-house', was compared to Venus:

> In pink and in white the sweet girl was drest
> Like Venus she look'd I declare
> And as June when drawn by her peacock's so gay,
> She sail'd in the regions of air
>
> For beauty our ladies have ever been famed,
> And for courage not carry the day
> So let's drink to the lass in an air ballon,
> Who look'd like the goddess of May.[122]

Simonet definitely played to the crowd, waving her handkerchief and then flinging it down into the crowd watching. Some men and women even danced through the air, as when four couples joined Blanchard in a launch during which they danced a quadrille, at least until some of the people were taken ill.[123] By playing to the interests of the crowd Blanchard became heroic in the eyes of some women, something noted in a poem written in his honour in which the author notes 'when you rose, impell'd by fear / Each bosom heav'd a thousand sighs; / To you each female lent a tear, / and held the 'kerchief to her eyes'.[124]

Other aeronauts also were connected with some modicum of impropriety during their launches. The *Morning Post* reported a wager, on the part of Lord Cholmondley, who claimed he could not only ascend in a balloon within twelve months, but that he would also take with him 'a lady on an amorous expedition'.[125] There is no record of whether or not he succeeded. Zambeccari, in a slightly backhanded manner, was accused of greed rather than immorality by accepting 300 guineas to take 'an old woman' on a voyage rather than the 'beautiful female volunteer who offered to ascend with him'.[126] More generally, a broadside, with a picture of woman with a large, balloon-like chest appeared with an accompanying poem saying the woman excelled 'in the inflammable air' and could never fall 'with her balloons'.[127] Similarly, a failed launch was described in a poem, 'the Lunar Travellers [*sic*]', in which the globe was 'flaccid and weak' to which the author added 'And 'tis not the first time, / From old age and cold clime, / A maid disappointed has been'.[128] If the balloon was not associated with sexuality, the gas inside could be. In some couplets addressed to one of the Montgolfier brothers the author writes the 'fair sex' has a 'tempting gas'.[129]

Lunardi, however, served as the most frequent target for licentious discussions about the relationship between ballooning and sex. The *Morning Herald and Daily Advertiser* called him a 'charming feller!' and claimed he walked among his admirers 'while his all surveying eye rolled over the adoring women with a conscious dignity, as much as to say '*Je suis Lunardi! – prenez garde de vos coeurs*!'[130] Interestingly, this British newspaper had the Italian Lunardi speaking French, associated at the time with sexuality and pornography. Lunardi also inspired women's wear – namely bonnets and garters – which, in turn, inspired people to poke fun at him. A somewhat salubrious poem written 'On the Lunardi Bonnet and Garters Worm by the Ladies' suggests:

> From the head of each Fair, down so low as the knee,
> Thy dominions, Lunardi are fix'd _____.
> Not a monarch so riche, or so happy can be,
> Since there's nought but an Eden betwixt![131]

The suggestion, that Lunardi was master of the air balloon as well as the space between women's legs, was commented on again and again. Certainly Lunardi seemed quite popular with the ladies. A popular science demonstration by Giuseppe Pinetti was usurped by the arrival of Lunardi who 'entered the house, just after the curtain was drawn up, with a party of Ladies, apparently vying with each other in shewing him attention and respect'.[132] Even his launches led to sexualized and scatological humour. A ascension in 1785 caused many people to crowd around the launch site including up the trees. This included members of the 'tender sex' there 'to see Mr. Lunardi rise with his balloon'. The anonymous author goes on to comment that the women had to wait in the tree for several hours leading to complaints from the people below of the 'temporary showers that descended from the boughs!'[133]

For a woman to travel alone with Lunardi in his balloon was certainly cause for comment. An article in the *Morning Post* suggested a certain 'Mrs. C— sw—y' had asked Lunardi to take her with him in his next flight 'determined to encounter dangers, hunger, and every inclemency of the sky, rather than not be talked of'. The article suggested, however, that whether this course of action was laudable or not, it would certain make her a focus of gossip. 'We doubt not', the author wrote, 'if she should faint away in the *manly arms* of *Mr. Lunardi*, when out of sight of this world, it would be for no other end than to be talked of'.[134] More crudely, the *Morning Herald* commented on Lunardi's reception by Mr and Mrs Nesbit. Quoting Lunardi, the article claims that Mr Nesbit was apparently out of the house when Lunardi arrived but 'his charming lady had prepared warm blankets for his reception, and that he was hardly stripped when Mr. Nesbit returned'. The article goes to say 'that Mrs. Nesbit is as chaste as the *moon*: and you shall never persuade us, Signor Lunardi, that you were the *man in the moon*!'[135] It was surely this reputation that led Lunardi, after one of his Scottish flights, to be honoured with membership in the local Beggar's Benison, a men's club devoted to sexual libertinism.[136] Lunardi even comments that, at one of his receptions in Scotland, 'he found the Town Hall full of Ladies' who vied with each other to help him repair his balloon. To thank them, he cut off small pieces of the balloon to give to his helpers who rewarded him with 'the most animated EXPRESSIONS of REGARD;' some of the women promised to set the pieces in lockets and wear them 'in remembrance of the event which introduced [Lunardi] to their acquaintance'.[137] Lunardi often received such attentions from women although they did not always manage to say the right thing. A lady who encountered Lunardi at the Pantheon where he was displaying his balloon told Lunardi 'that she saw his *bagatelle rise* (meaning the balloon) with pleasure, for it proved him a *man*, and much superior to *Blanchard* in that way'.[138] Certainly the innuendo, of a well-endowed Lunardi, was intentional on the part of the author, if not on the part of the lady.

Class

Commentators frequently mentioned the various ways in which individuals from different social backgrounds reacted to the appearance or launch of a balloon. James Gennittie, for example, launched a balloon in late 1783 from London. This small globe journeyed as far as Leominster, a town about 136 miles north/northwest of London where many of the inhabitants were unprepared for the sight. A local man, named Benjamin Fellows wrote a letter back to Gennittie saying that the local citizenry 'were this afternoon much alarmed by the appearance of a large body of a globular form floating in the air; the fears and consternations that seized the minds of the gaping spectators upon this extraordinary occasion are beyond conception'. The people in this small town, however, were eventually 'relieved from the dreadful apprehensions they had formed' when the balloon alit nearby and they found the note Gennittie had attached explaining his experiment.[139] 'Fear' and 'consternation' in the 'gaping' faces of the people offers a thumbnail sketch of how educated people viewed the general population's understanding (or lack of understanding) regarding science in general and ballooning in particular. These themes were repeated throughout the early age of ballooning and throughout Europe. Certainly, the response of the villagers of Gonesse, who attacked the balloon which landed in their midst, encouraged people to ridicule lower-class responses to aeronautics (see Figure 3.2)

Many descriptions of the audience commented on the social diversity of the people who viewed launches, the 'idlers, thieves, Pickpockets, Princes royal, Nobles, etc.' who all came out to watch.[140] James Deeker, for example, who oversaw several launches in the north of England, drew 'an astonishing multitude of all ranks and descriptions of people from York and the neighbourhood, as well as from various distant parts of the country'.[141] An anonymous observer commented on a 1784 ascension from Grosvenor Square in London noted the large number of people gathered at nearby Hyde Park to see the balloon rise. This audience was composed of 'noble as well as ignoble visitants'.[142] To put it another way, launches drew 'the attention of both the wise and vulgar'.[143] Such a range was fairly common and, occasionally, something about which people commented. Thus, in the *Gentleman's Magazine*, the author of a short article on an ascension noted it 'drew an innumerable multitude, of all ranks, together'. The author added 'it was really curious to listen to the discourse, and to observe the different traits of so many vacant faces, who, though assembled, had no conception of what they came to see, or perhaps, more properly speaking, what they had pronounced beforehand was impossible to be seen'. He broke his observation down into several categories arguing 'the notions and opinions of this motley multitude were certainly as various as their situations in life were different'. Thus, 'the populace', which in his view 'composed the far greater part of the company', felt that the balloon was doomed to failure since 'no Christian could fly through the sky'. Here, the author gives the majority of the people a religiously slanted view implicitly claiming the average person equated balloon flight with witchcraft. The balloon could not fly during the day because 'Goblins and Spirits were not permitted to ramble abroad till the dead hour of the night'. The next class, presumable artisans, 'had very little more faith than their fellows'. Words are placed into the mouths of the workers; they say 'they could not think as how it could be that a bubble could carry a man'.[144]

After dispensing with the bulk of the population, the author turned next to 'the middle ranks' who were 'doubtful, but not without hope'. After this, he turns away from describing the people by rank or class and turns to their intelligence. 'The enlightened', he argued, 'were anxious for the event, and were not without sharing in that concern which every sensible mind could not but feel for the issue of so hazardous an enterprize'. This was followed by another group, 'men of real science', who were 'otherwise affected' and 'could not help expressing, by their looks, the sympathetic concern they entertained, lest some untoward circumstance should intervene, to defeat, or even to delay the execution' of the launch. The article ends by returning to the use of rank; all the people mentioned heretofore are described as 'the subordinate classes'. Now, the author turns to 'those of elevated rank who look upon the life or death of an individual, with equal indifference, and the good or ill success of an experiment, with equal indifference, and calculated only to kill time, and contribute to their amusement'.[145] These people considered the balloonist, Lunardi in this case, to be another Bottle Conjuror, trying to dupe the people and rid them of their money. The article, then, lauded the men of science and the enlightened while everybody else, of low or high rank, seemed to be missing the point entirely.

Certainly many members of the audience came from the aristocratic elite, including the monarchs of Europe. Richard Crosbie sent off a balloon in Dublin in 1784 in front of 'some hundreds of Ladies and Gentlemen, many of them of the first distinction'.[146] Apparently many of the audience members at Lunardi's ascension from the Artillery Grounds 'were of the genteeler sort'. These people, however, were the paying customers. The non-paying spectators were described in the same article as a 'mob' who were 'a good deal exasperated' by the short duration of the flight. Lunardi, the author of the article claimed, 'escaped from their fury' with some difficulty.[147] Nonetheless, monarchs and nobles frequently appeared at launches and, occasionally, funded them. Louis XVI, along with most of the royal family, attended the event held at Versailles in 1783. One commentator wrote that Louis and Marie Antoinette even examined the machine although they quickly retreated. At that time the Montgolfier brothers were still experimenting with different methods of creating the hot air that filled their balloons; at Versailles they apparently utilized 'damp straw' together 'with pieces

Figure 3.2

of decomposed meat' which produced, along with hot air, a rather disturbing odour.[148] In Britain Aimé Argand, a Swiss savant, demonstrated a small balloon for George III and his family at Windsor Castle. Lunardi also performed for the British king who, along with the Prince of Wales, William Pitt, and other ministers of the king went to watch Lunardi, the king remarking 'we may resume our deliberations on the subject before us at pleasure, but we may never see poor Lunardi again'.[149] The Duke of Cambridge and the Princess of Gloucester watched one of Sadler's ascensions in 1814 with a particular interest as to the inflation of the balloon.[150] In The Hague, the balloonist Diller performed a launch in the presence of the Princesses of Orange and Nassau, the latter of which 'consented to cut the string' initiating the flight.[151] More generally, an aeronaut in Paris dedicated his newly designed globe 'to the nobility', as well as to the clergy, of France.[152]

In addition to references to the nobility commentators also mentioned social status when talking about workers and criminals. These allusions were often less than positive as when it was proposed that 'an intelligent worker' be sent up in the first manned hot-air balloon rather than risk the life of a savant.[153] While this idea was ultimately overruled – too many savants were vying to be the first person aloft and were unlikely to cede the honour to a worker – it remains a telling bit of social commentary. Similarly, the idea of sending up balloons populated with convicts demonstrates the place certain members of society held during this period. James Watt, in a letter to Joseph Black, described one case where a balloon was launched with '4 or 5 Criminals tyed [*sic*] to it' although it is unclear if this launch ever really took place.[154] In 1783, before anyone else had ascended in a balloon, the *Pennsylvania Packet* claimed 'two criminals, condemned to be broke upon the wheel, have petitioned to be sent up in some kind of cage to be fastened to a globe, on condition that if they get down alive they should have their pardon'. The author of the article suggested that if they were allowed to go, then they would have been fortunate since to die in the air 'as it is most probably they must' would be a better death than the wheel. In any case, their request was denied.[155] The same newspaper later reported that the French were considering sending convicts aloft 'with promise of their lives and pardon if they escape tumbling on some church leads'.[156] Convicts and workers clearly operated near the bottom of the social scale.

Peasants, however, dominated the lower ranks and came in for considerable ridicule and scorn when it came to balloons. They were almost uniformly treated as childlike and superstitious with respect to aeronautics. Blanchard, for example, reported that the peasants who saw him and his companion Boby land asked 'Are you men, or Gods? What are you? Make yourselves known'. Blanchard replied that they were men and, to prove it, threw their coats out of the gondola. The peasants 'seized them eagerly, and began to divide them in pieces', a scene which 'afforded us great amusement'.[157] Blanchard clearly felt superior to the peasants among whom he found himself and openly laughed at their response to his arrival. Some peasants in Spain had a religious reaction to the arrival of a balloon. When Lunardi landed at Horcajo the villagers 'took him for a saint come down from heaven and carried him in triumph to their church'.[158] As belief in religion in the age of Enlightenment was viewed askance by some savants, the association of peasant attitudes toward aeronautics with excessive devoutness was implicitly critical. When Lunardi landed after one of his Scottish voyages, a nearby peasant couple, 'not having ever seen a balloon, and observing it fall rapidly with uncommon grandeur from the clouds, their emotions of mind were on this occasion inexpressible; and they both, as it were by instinct, fell down on their faces, and continued in this groveling position for a few moments'. Eventually, the man, 'encouraged by his religious faith', stood up, 'incited his wife to follow the example' and exclaimed they were going home to pray for their salvation believing the balloon to be evidence of the Day of Judgement.[159] The attitudes of people like those encountered by Blanchard were surely exacerbated by the infamous response of the villagers in Gonesse to the arrival of the first hot-air balloon released from Paris. They mistook the balloon for a 'ferocious beast' and attacked it although some of the locals did, apparently, recognize their error.[160]

While some peasants saw globes as descending from heaven, others apparently viewed them as Satan's work. One of Lunardi's flights in Britain ended with him shouting to some peasants in a field to come grab the rope he had thrown out. They responding by saying 'they would have nothing to do with one who came in the Devil's house, or on the Devil's horse (I could not distinguish which of the phrases they used) and no intreaties [sic] could prevail on them to approach me'.[161] In this instance Lunardi had trouble with the language skills of the peasants and critiqued their speech patterns. An account of an Irish launch described the local peasants as 'terrified at so extraordinary a phenomenon, thinking it was either the devil' or some kind of bird.[162] One peasant, who knew nothing of balloons received Blanchard 'in a very rude manner' and 'thinking that he must be in a league with the devil, attempted to stab him with a knife'.[163] Blanchard survived the assault although attempts to kill him seemed to occur frequently. A peasant apparently fired a gun at him in 1784 imagining he was some 'strange animal'.[164] A servant working for a 'Mr. Emery, Farmer', had a balloon land at his feet. When he tried to grab it, 'it immediately jumped over his head' scaring him. He 'ran to the house, exclaiming that he had seen a dancing devil in the shape of a large meat pudding'. The other servants ran to his rescue, armed with 'prongs, rakes, &c'. and found the 'supposed daemon in a state of perfect tranquility'. They subsequently examined the balloon and found a letter attached explaining the origins of the balloon. The man who first found it then claimed the right to exhibit the balloon in town, something the others agreed to 'as a compensation for the poor fellow's fright'.[165] Lunardi reported that when a Scottish woman saw him land she 'touched his clothes, and his body' and, after rubbing her eyes, declared 'I am sure there is nae Glammary here, but oh, Sirs, it's a fair pity he should be a Papist'.[166]

Charles also had a somewhat demeaning attitude toward the peasants. He described them chasing after his balloon 'like children following butterflies across the fields'.[167] Here, instead of overweening ignorance the peasants are simply childlike. Newspaper accounts almost always commented on the ignorance of the peasants. When Blanchard landed near Philadelphia it was 'to the great alarm of the peasants who were reaping'; another balloon in America elicited 'the wonder of the laboring People'; a farmer 'was so terrified, that taking the aeronaut for some sky-man' he ran to his house to get a gun so he could defend himself.[168] A flight by Garnerin in Britain 'excited the utmost astonishment among the country people'.[169] Another voyage caused fear in the 'cattle, cows, sheep, dogs, [and] ducks' while all the villagers cried 'papa' and 'mama'.[170] Perhaps most humiliatingly, one author described how a chemist named Nicolas, who launched a balloon from Nancy, France, went out in search of it 'asking the peasants he encountered if they had not seen a balloon?' One of the peasants responded to this query 'in his patois, "Je n'qenaichonme les Ballons, j'ons vus un graux ougés, & lou Compere a dit qu'ç'ato mairque de fraux"'. The author, of course, then deigned to translate the peasant's comments into proper French writing 'Nous ne connaissons pas les Ballons, nous avons vu un gros oiseau, & le Compere a dit que c'était marquee de froid.'[171]

In some cases the peasants were described less derogatorily. Thus, the peasants watching the launch of the Robert brothers' balloon simply cried out 'Long Live Robert!'[172] Other peasants sought to make money from their encounters with balloons. A peasant from Zevenhuysen, in the Netherlands, tried to hold one of Blanchard's balloons hostage. He argued that anything that fell from the sky and landed on his property should belong to him.[173] Similarly, a farmer in Sussex who discovered a balloon on his land 'unlike the terror stricken peasants of Gonesse, did not destroy the balloon but bore it away to his barn where he exhibited it to wondering villagers at a penny a peep'.[174] A farmer near Epping in Britain found a balloon on his land and returned it to the owner for a reward of one guinea.[175]

The Unruly Audience

Not all descriptions of the audiences for balloons were positive.[176] Whether they described the audience in general terms, or referred to the specific audience which showed up for particular launches, some authors made it clear that the people interested in balloons were potentially dangerous and riotous. In Boston, for example, the *Independent Ledger* noted that 'the *Balloon rage* increases rapidly in this town'.[177] Such a violent modifier would seem to indicate a certain inability to exert self-control on the part of those persons interested in aeronautics. Even when rioting was not likely, some balloonists were less interested in having an audience for their work. Thomas Baldwin tried to keep the crowd at bay by urging the audience to follow strict rules. 'In order to prevent an Interruption of the Process in the Inflation of the Balloon', he wrote, 'no Persons were to be admitted WITHIN THE CIRCLE, except those Gentlemen who politely undertook it IN TURN to hold the Lines which detained the Balloon'.[178] This emphasis and insistence on politeness reinforces the omnipresent possibility of its opposite, namely riot.

Certainly, the audiences that showed up for launches were potentially unruly in a variety of ways, including the possibility of riots but also other sorts of behaviour. The 'bustle and confusion which prevailed in the Artillery ground' at Lunardi's launch engendered 'a general cry of pick pockets!' until the Prince of Wales showed up and things calmed down.[179] The same was true of the launch by Mr Astley in 1784, an ascension 'many of the spectators will have reason to remember' since 'a more ample harvest for the pickpockets never was presented. Some noblemen, and gentlemen lost their watches and many their purses'.[180] Simon-Nicolas-Henri Linguet suggested that 'balloons seemed a suitable tool for facilitating larcenies and abductions'.[181] A broadside showing a woman flying out of a house suggested that balloons could have 'dangerous uses' and that the 'perfidious globe' was used by the woman to steal a dress.[182] Such possibilities, even if they rarely occurred, led some to critique aeronautics in general. One author claimed that 'profligacy and idleness' were 'promoted by every balloon that rises in the air'.[183] A letter printed in the *Salem Gazette* and written by a 'Lay Preacher' claimed,

> Better it is for thee, O son of riches, to succour genius, to relieve the miserable, and shelter the naked, than to enjoy the glorious privilege of beholding a Balloon inflated. Gratification of this idle curiosity will, like the volatile machine its source, vanish into 'air, into thin air;' and leave not a remnant of complacency behind.[184]

Clearly, some people associated the growing interest in ballooning with an attack on society and, in this latter case, on religious morality. In a few cases, this caution was not without merit.

Riots typically occurred due to the inability of the balloonist to launch the balloon whether or not they were at fault. Thus, when the wind was blowing too hard for a flight in May 1784 in Bordeaux, causing the aeronauts to keep the doors of the garden, where the balloon was to be launched, closed, there were great 'grumblings on the part of the multitude', and especially the 'workers angered at having lost their days' work'. As a result, the mob 'destroyed the barrack from which the tickets were distributed, broke the windows of a neighboring house, and forced open the doors of the garden'.[185] Commentators often noted the dangers of ballooning. Benjamin Franklin wrote:

> It is a serious thing to draw out from their Affairs all the Inhabitants of a great City & its Environs, and a Disappointment makes them angry. At Bourdeaux lately, a Person who pretended to send up a Balloon & had received Money of the many People, not being able to make it rise, the Populace were so exasperated that they pull'd down his House, and had like to have kill'd him.[186]

As Franklin notes, raising the expectations of the people, without also raising a balloon, could lead to severe consequences. Guards or police often were called upon to quell a riot. After a failed launch by Zambeccari in 1785 the mob 'tore down the scaffolding' and 'reduced part of the wall to ruins' before the guards arrived and 'fourteen of the riots were taken into custody'.[187] In 1812 Jacob Degen's launch necessitated 'the intervention of the police' to prevent disorder after a disastrous attempt to launch a balloon from the Champ de Mars.[188] As a result, aeronauts frequently had recourse to hire guards 'to maintain order'.[189] Even female aeronauts did not escape from the rancour of the public. Eliza Garnerin failed in a launch in Madrid and 'narrowly escaped with her life'; armed guards had to smuggle her away and hide in the local prison 'to save her from the public indignation'.[190]

Many aeronauts expressed concern over the possibility of unruly behaviour on the part of their audience. Lunardi expressed concern before his planned ascensions in Scotland. 'I have requested the attendance of the military' he wrote, adding 'the fear of disappointing the public curiosity, which is now highly raised, has outweighed every other consideration'.[191] A Mr Michel reported to the people of Philadelphia the circumstances of his failed launch. When an accident caused the gas to escape from the globe, the crowd got angry. They 'began to tear up the fence, then rushing in, demolished the balloon and parachute'. They also took the strongbox with the gate receipts and drank all the liquor at two nearby bars.[192]

The response to failed launches could be severe. The audience frequently took their frustrations out on the balloon and the surrounding equipment. Although launch sites were frequently divided between paying and non-paying customers, the riots tended to bridge this gap with the non-paying audience tearing down barriers between the different sections. It is unclear who exactly participated in the riots although the upper-class commentators typically assumed it was the lower-class members of the audience who did all the damage.[193] Whoever participated, the rioters tended primarily to vent their anger by tearing the balloon into shreds. In Holland, a globe 'on which was beautifully painted an Eagle with a Thunderbolt in his talons, was torn to pieces by the disappointed crowd' while the would-be aeronaut jumped 'into a cellar, to avoid the resentment of the populace'.[194] Occasionally, the mob would act out the launch on their own, thus taking charge of the launch process, and usurping the role of savant for themselves. Thus, after Tytler failed to launch a balloon in Edinburgh, the crowd of spectators 'seized the basket, and after carrying it through the streets, in a sort of triumph, at least burnt it to ashes'.[195] This, in fact, mirrored the path of a launch since the end of a balloon trip often ended with the balloon being damaged and, occasionally, catching fire. The willingness of the audience to take control of their own spectacle demonstrates the mental attitude of the audience toward this particular event. If the theatrical production could not manage to succeed on its own, they simply took over the process and entertained themselves.[196]

*****

The social breadth of audiences, and the balloonists, clearly reveals how widely aeronautics entered into popular culture in the late eighteenth and early nineteenth centuries. People from all walks of life attended launches; people from many walks of life launched globes. This did not, of course, make everyone happy and the lower classes in particular came under some criticism for their interests. Women also attended frequently, and also launched their own balloons. While this led to some amount of sexual innuendo and criticism based on immorality, for the most part women were accepted members of the audience and the group of aeronauts. One reason for this may have been due to their purchasing power and their willingness to buy tickets and aid subscriptions. As the consumer society developed during this time period, women were identified as an important component of those with money and leisure to purchase access to science.

4 CONTROLLING THE SKIES: STATES AND BALLOONS

At Napoleon's coronation, André-Jacques Garnerin, then imperial aeronaut to the new monarch, launched several balloons to honour the occasion. This was also a method of announcing the event since the balloons contained news of the Emperor's crowning and would carry that information to wherever the wind took them. The largest of these balloons sailed across the Alps and landed in Italy, near Rome. Although almost certainly an apocryphal tale, the balloon reportedly descended and touched ground first at the tomb of Nero, one of Rome's most infamous emperors. The king of Italy gave the balloon as a gift to the pope, who had been present at the coronation (somewhat redundantly since infamously Napoleon placed the crown on his own head). The pope duly deposited the balloon in the Vatican with an inscription 'which records the wonderful rapidity of the passage, and the solemn occasion on which it was launched from Paris, as well as the success of M. Garnerin, who constructed and launched it'.[1] Nonetheless, Napoleon expressed his displeasure at the landing site of the balloon, true or not, and fired Garnerin replacing him with Marie-Madeleine-Sophie Blanchard, widow of the prolific aeronaut Jean-Pierre Blanchard.[2] The political uses of ballooning stretched between two emperors, one of whom clearly resented his association with the other.

Almost from the moment of their invention, balloons drew enormous crowds eager to watch launches and see for themselves how humans had conquered the heavens. At times, people travelled from miles around to see a launch; they took time off from their work, collected in large numbers near launch sites, climbed on top of buildings and took over all available spaces. Aeronautical experiments proved wildly popular, much to the advantage of those using subscriptions to fund their work but much to the chagrin of savants, who hoped to use balloons for scientific research, and the state, which saw in launches a greater potential for riot and misrule than they did for economic gain or scientific expansion. In all cases, however, individuals faced difficulties in controlling access to launches. Aeronauts funded most flights through subscription, which meant large crowds were bound to gather. In addition, once the balloon reached a certain height anyone could witness the experiments whether or not they had paid for the privilege. This helps explain the importance of controlling access to the workshops where aeronauts manufactured their balloons and the fields and gardens where the launches often took place. It also helps explain the desire of the local and state authorities to police the practice of ballooning.

This chapter explores the myriad methods used to control access to launches and analyse the reasons why different groups, including savants, state authorities and balloonists, tried to set limits on who could attend ascensions and when it was acceptable to conduct experiments. While the state certainly allowed the use of globes for state functions, it wished to control their availability elsewhere. In addition, this chapter examines what occurred when the state's worst fears were realized and a riot erupted. Part of this story includes a discussion of the different attitudes of balloonists and others to the various social groups who viewed launches. Since the audience might perform the task of witnessing, helping fund the launch, or, from the negative side, rioting and destroying the balloon if the launch did not go as planned, the balloonists' relationship with his or her audience was a crucial and tense zone of scientific popularization.

Fêtes

Balloons figured prominently in state festivals after 1783. From the moment when the Montgolfier brothers first demonstrated their new invention before Louis XVI, Marie-Antoinette and the rest of the nobility at Versailles, balloons became a common feature of elite entertainments. If nothing else, nobles were interested in aeronautics; the Prince of Wales, for example, attended one of Vincenzo Lunardi's launches in Chelsea. For France, the historian Jacques Godechot claims 'by the end of the Directory balloons no longer seemed to serve as an amusement or a surprise for the masses'. Instead, he suggests, balloons had taken on a political function and had appeared at each Bastille Day celebration since 1795.[3] Certainly, aeronauts festooned their balloons with patriotic symbols. A 1798 launch in France included a globe ornamented with 'garlands and patriotic inscriptions'.[4] The Union of Piedmont with France in 1802 included a ceremony punctuated with 'elegant and pleasing' fireworks announced 'by the ascension of a balloon'. The celebrants followed this with dancing.[5] During the French Revolution, balloons frequently appeared in celebrations, especially of major events including Bastille Day. In December 1789 an announcement in the *Journal de Paris* for an ascension from the pleasure gardens of the Ruggieri brothers – most well known for their fireworks displays – suggested that this balloon should also be used for the upcoming Bastille Day celebrations in July. The aeronaut in charge of this launch, who remained anonymous, proposed this balloon 'in honour of the patriotic citizens' who had died during the taking of the Bastille. This event would not be free, of course, and those who wished to attend could purchase a prospectus describing the celebration. This prospectus also served as the entrance ticket.[6] An 1801 Bastille Day festival included 'pantomimes, different theatrical entertainments, and concerts' along with 'the ascension of balloons'.[7]

Not all Bastille Day launches went well. In July 1790 the French state sponsored a number of events in Paris in honour of the Festival of Federation including an ascension from the Champ de Mars. The aeronaut, however, failed to inflate the globe sufficiently to take his weight and he sped off into the crowd leaving the audience free to try and launch the balloon themselves. Writing in an American newspaper, one anonymous commentator noted 'the surrounding spectators made ineffectual efforts to send it up without any weight' but 'it would not do, and the balloon, after ascending a few feet ... fell upon the heads of the people in the amphitheatre'. The author went on to criticize the ignorance of the 'country confederates and their spouses', by which he meant the rural officials elected to the National Assembly. These country bumpkins apparently had heard of 'balloons with inflammable air' but 'were afraid it might burn them'. Thus, they 'jumped over the seats and balustrade' and risked 'breaking their necks for fear of an accident'. Other, more daring, individuals tried to complete the task of filling the balloon by burning some straw nearby. The result 'was an explosion that scorched five or six of them so miserably, that for the rest of the lives they will not be likely to make any more experiments'.[8] Thus, the failure of this particular launch allowed the commentator to implicitly critique the intelligence and capabilities of the new leaders of France.

The French had a rival discussion of the same event in which the balloonists launched the globe, decorated with 'the colors of the nation', to the sound of military music and amid spectators all giving homage to the country. The aeronauts planned to go and see if the inhabitants of the moon were free and, if not, help them create their own declaration of the rights of man and thus overthrow tyranny. In fact, the pamphlet discussing this launch explicitly compared the Seine River to the Tiber River in Rome, a reference meant to elicit a link between the old Roman Republic and the new French state.[9] In the short, anonymous pamphlet *Fête aérostatique*, the author wrote that using balloons, which the French had the honor of discovering and perfecting, was one of the best ways to celebrate the nation.[10]

A proposed aerostatic experiment during the Festival of the Federation in 1798 allowed the French to play out their frustrations against the British. A small, tethered balloon rose above the Champ de Mars with two aeronauts aboard. Meanwhile, a small building resembling an English warship was erected in the middle of the field. The aeronauts then shot bullets, of a 'special composition', which would ignite the building'.[11] This celebration thus allowed the

French to attack the British metaphorically, even if their actual ability to deal with the British blockade of France was somewhat less effective.

Some launches during the French Revolution were not linked specifically to national festivals but still had political overtones. B. Lallemand de Saint-Croix undertook an ascension in September 1791. His balloon was decorated with patriotic images representing liberty, love of country, France and the law.[12] Standing in the gondola, shaped like a rooster, he held the new constitution in his hand while sailing over the Tuileries, the Louvre and the Champs-Elysées all to the accompaniment of the applause of the immense crowd that had gathered to watch. He threw out copies of the constitution to the people below.[13] Garnerin, along with his family of balloonists, conducted several launches during the revolutionary era including the ill-fated launch at Napoleon's coronation mentioned at the head of this chapter. More successfully, in 1798 Garnerin ascended, with 'a young woman' from the Parc de Mousseaux amid a row of statues and vases, from which one could smell perfumes from Asia. The elegant decorations, Garnerin claimed, rivaled, 'in good taste and in magnificence, those which were given at Versailles, under Louis XIV'.[14] This is an interesting claim to make; clearly the rampant republicanism of the early part of the Revolution had relaxed allowing an appropriation of some royalist-inspired spectacles. Before one of Garnerin's balloons reportedly landed on the tomb of Nero, viewers also attested to the great success of his launches for Napoleon's coronation. The *Moniteur Universel* reported:

> Yesterday afternoon five balloons were sent up from the Place de la Concorde. Their departure was preceded by that of a small air balloon which rose with great speed and appeared in the sky as a shining point of light, coloured by all the rays of the sun. The largest of the five balloons, which was of a most majestic size, carried an eagle, the wings of which were outspread and which carried in its claws two large flags bearing the name of Napoleon, the Emperor. The give balloons rose up to a great height and burst into flames in a most spectacular manner; the eagle came down to earth, borne up by its wings and by the flags which it carried.[15]

The intentionally exploded balloon showed the power of the balloonist, and by association the emperor, over an object in the heavens. In general, this use of aeronautics at the coronation served, initially, to heighten the status of the new emperor. It was only later that Garnerin fell out of favour.

When Napoleon sat in exile on the island of St Helena, some commentators thought he might make further use of aeronautics by taking flight in a balloon to escape his fate.[16] Certainly, the rest of Europe continued to entertain themselves with globes. During the Congress of Vienna, an aeronaut named Kraskowitz, described by one attendee as a rival to Garnerin and Blanchard, launched a balloon 'of enormous dimensions' from which he unfurled 'a great number of banners with the colors of the nations represented in Vienna'. The Prince de

Ligne reportedly suggested that if Kraskowitz was a sceptic, he had an opportunity for seeing just how small the members of the Congress really were as he soared above them. 'All these prominent persons, who appeared so little when viewed from above with an untrammeled eye', he supposedly opined, 'must lose a great deal of their importance'. If a gust of wind took him even higher, then soon his gaze would see only 'indefinite masses, a chaos without names'.[17] While this hyperbole remains unsubstantiated, balloons were used to entertain the members of the Congress of Vienna and their entourages. When Louis XVIII took over as king in France one of the celebrations included Marie-Madéleine Sophie Blanchard who descended in a balloon decorated with the royalist fleur-de-lis.[18]

Politics, Balloons and Satire

Balloons served a variety of political purposes at both the national and international level. In some instances, a launch might lend itself to potential international cooperation. Blanchard certainly hoped for this when he planned his journey across the English Channel. The *Morning Post* reported that Blanchard trusted 'that the vessels who may happen to see him in his voyage, whether French or English, will give him timely assistance in case he should come down'.[19] In most cases, however, balloonists fed national pride rather than promoted mutual aid. One British newspaper, noted that several foreigners – namely Lunardi, Blanchard and Zambeccari – planned 'aerial excursions' leaving it to the 'ingenious English aerial traveler, Mr. Sadler' to 'make another effort for the honour of his country'. The article added that 'Mr. Sadler has already soared higher than either Lunardi or Blanchard, and to the honour of Old England, is the only person who has been his own projector and chymist in this or any other country'.[20] Since the first aerial voyages in England were conducted by Italians, Sadler's successes at least gave the British some sense of victory The constant battles between Great Britain and France also led to a sense of contest over ballooning. Lunardi claimed that those two nations watched each other with great 'anxiety and jealousy' especially with respect to discoveries 'in science or an improvement in fine arts'.[21] It did not help England's sense of self-esteem that French inventors had developed ballooning first. As Louis Sébastien Mercier noted, that invention would 'immortalize' the reign of Louis XVI.[22] Louis even sent twelve taffeta covered globes as a gift to the emperor of China.[23] This helps explain why the British even accepted the triumph of Lunardi, at least when compared to French aeronauts such as Blanchard. A popular song claimed 'that Blanchard's been puffing and puffing again' in vain, so 'Lunardi's balloon, the English Balloon, Lunardi's balloon for me'.[24]

As Mercier suggested, the French received the greater share of glory for this invention. Even some people outside France acknowledged the priority of

French claims in this instance. Thomas Baldwin wrote that while Cavendish had developed hydrogen, no one would have thought to apply it to balloons had not the Montgolfier brothers invented their globe. Thus, he concluded, 'in this the French are still before the English'. He goes on to claim that 'half that Time which men of Fortune in France dedicate to Taste, Invention, and Refinement; Britons spend among the Beasts and Birds: the other half, at the Bottle and in political Cabals'.[25] In spite of Baldwin's assertion, the British and French often put forth rival claims for priority in the invention of ballooning.[26] At other times commentators suggested that while the French may have invented aeronautics, it was the British who theorized it. Benjamin Franklin lamented the role played by 'national jealousy' in preventing the English from immediately pursuing ballooning themselves since, he felt, 'in their Hands it might have made a more rapid Progress towards Perfection, & all the Utility it is capable of affording'. Although Benjamin Franklin erred in thinking the British could find a use for balloons any faster than anyone else, he accurately portrayed the hesitancy with which the British turned their attention to a foreign invention. Eventually, of course, people from all over Europe sought to honour their cities through launches. There was even a hint of colonialism in battles over ballooning. A poem printed in a Massachusetts newspaper claimed 'And Since Man's selfish race demands / More empire than the seas or lands; / For him my courage mounts the skies, / Invoking Nature whilst I rise'.[27] This hypothetical use of balloons to extend the domains of states into the heavens also found expression in the writings of Simon-Nicolas-Henri Linguet who suggested that the Hottentot's would think the French were gods if they demonstrated a balloon in Africa.[28]

Balloons also found use as a metaphor for political actions, mostly in a negative sense. Aerostatic globes, full of hot air and unable to be directed, provided ample fodder for those seeking to deride their political opponents or offer a critique of the state. An epigram from the *Bath Chronicle*, for example, noted that 'To fill a Balloon with the lightest Phlogiston, / A rustic Philosopher try'd, and oft miss'd-on" / To the Parliament House', says old Sly-Boots, 'repair"; / You'll there find enough of Inflammable Air'.[29] While the distance between hot-air balloons and politicians being full of hot-air is not much of a conceptual leap, it still offers a decent attack on the affairs of state. In another instance, ballooning found its way into a critique of voting practices. In this case, a song conveys how a woman used her own personal supply of gas, expelled from her bottom and contained by sewing up her petticoat, to fly about the countryside and transport voters from one place to another. 'Thank God', one of the voters cries, 'I am an outside passenger'.[30] An American newspaper critiqued the French state during the revolution by claiming that a launch by Garnerin in the company of a young woman ended badly when they landed outside Paris and were confronted by a local official who 'asked them how they came to travel in the air without a passport' and subsequently took them into custody as emigrants endeavoring to escape in a balloon. Only the arrival of people from Paris, who had been following the globe, convinced the magistrate to release them.[31] Similarly, another American newspaper satirized a launch by Garnerin by claiming 'the physician could not mount, the barometer of the Jacobins being too low, and their gaz not able to raise them to the height of "the mountain".' These were explicit, and unfavourable, references to two of the major political parties of the French Revolution, the Jacobins and the Mountain. The author goes on to claim, erroneously, that after the failure of the ascension, the job 'destroyed the balloon, and the doctor [Garnerin] was arrested for the amount of the entrance-money (150,000 livres), which, after the real expenses are deducted, is to be applied to the relief of the poor'.[32]

Jacques Pierre Brissot de Warville claimed that while 'in France all dream of balloons, in London all are occupied with India and the ministry' and that the Montgolfiers 'in London are the Foxes and Pitts'.[33] Even if Brissot's caricature of the British preference for politics over ballooning were true, it still did not prevent contemporaries from littering their political commentary with aeronautical references. William Pitt the younger must have had mixed feelings about aeronautics; he frequently found balloons used against him. 'His enemies are obliged', writes the paper, 'to make his going to Brighthelmstone, his walking in the fields, or his talking to Mr. Steele, a capital crime. Then his going to see the Balloon launched in the Artillery-ground was boyish in a Prime Minister, and a sin never to be forgiven'.[34] Lunardi conducted this launch which also counted the Prince of Wales among the audience members. The same newspaper suggested that if Pitt 'should again dissolve the Parliament' it was probable that Lunardi might get elected for 'many of the principle cities, if not for the Universities'.[35] An American journal suggested that if Napoleon used a balloon fleet to invade Britain it would behoove Pitt to 'put the game-keepers and country gentlemen'on alert so they could prevent the attack by shooting the globes out of the sky.[36]

The *Morning Post* also used this new invention against politicians in general. One satirist suggested that the claim 'inflammable air is a new discovery' was unwarranted since 'have not many members of the Lower House gone up to the Upper House in balloons of this air?' and 'is it not the fashionable air in Ireland?'[37] The same author also posed the question of 'what but inflammable air produced the riots at Boston, and the subsequent American war? Was it not inflammable air which sent up the British cabinet as far as the lofty idea of conquering America?'[38] An essay discussing the aeronautical origins of the American Revolution never did appear.

Along with the British, the Americans also found uses for balloons. In 1787, to honour the birthday of George Washington, the city of Petersburg, Virginia hosted a dinner assembled at the home of a local resident, Robert Armistead, where people enjoyed dinner, singing, cannon fire and a launch 'which equally excited curiosity and applause'.[39] In Boston, a party in 1818 held to honour the anniversary of the American Independence was celebrated with a variety of 'rational entertainments' including the launching of two balloons.[40] These balloons were provided by A. Guerin, a seller of fine hats in south Boston who also advertised 'super-extra Beaver hats of the latest fashion and much cheaper than they can elsewhere be had' along with 'a complete assortment of Youth's and Children's Hats'.[41] Thus, as early as 1818, corporate, rather than state, sponsorship of the balloons maintained the public's control of aeronautics while also making strong connections between aeronautics and commerce. Ever earlier a Mr Martin, the owner of the Summer Theater in Philadelphia, used balloons as part of an Independence Day celebration, which also included fireworks. He played on the emotions and patriotism of the people by writing that he was so 'grateful for the many and often repeated favors he has received from the Ladies and Gentlemen of Philadelphia' that he planned his own celebration 'amid the other demonstrations of National Joy'. Martin charged fifty cents but pointed out that the 'Summer Theater is admirably well shaded; it is commodious and airy, and so contrived that the whole performance may be seen from any part of the Theatre'.[42] More simply, a man named Citizen Tronche, perhaps visiting from France, launched two balloons in Boston during January 1793. Two balloons, one twenty-five by twenty feet and the other nine by seven feet, had signs emblazoned with 'Liberty and Equality' on one side and on the other 'the Flag of the United States'.[43] Balloons, at least while waiting to be launched and for the first few minutes after the ascension, proved extremely useful as propaganda.

Authors also used balloons as a pessimistic illustration in American politics. One author, satirizing the expansion of government, proposed a new state model called 'Mammothocracy'. The author pointed to inventors of automata and perpetual motion devices, along with 'M. Blanchard, constructor of Air-Balloons' as sources of further information.[44] Associating Blanchard with automata and the problem of perpetual motion indicates that, to at least some people, ballooning was at best a marginal science with little use. A critic of John Quincy Adams's position on France in 1812 claimed 'never was a balloon inflated with inflammable gas more light or more showy, or less substantial' than Adams's letter arguing against closer ties to France.[45] This allusion draws in several elements: the author links ballooning with spectacle and entertainment, but certainly not seriousness, while at the same time connecting this frivolousness with France, where balloons originated. In addition, the inflammable air illustrates the hot-winded nature of Adams's statement. An 1806 accident at a launch by Louis Toussaint at Woodstock Green provided a forum for some political comment. At the moment of the ascension the balloon caught fire and 'was instantly consumed to ashes'. One observer suggested the audience learned a valuable moral – namely, 'that many things of the present day are perishable'. In addition, the destruction of the balloon 'was emblematical of the downfall of that Sham-Patriotism in our country, which dwindles to nothing before the glare of reason, and is as rotten as paper, itself'. The aeronaut, the author argued, did not burn his balloon on purpose just to provide this moral although he did suggest that Toussaint 'is like those self-made mushroom patriots, who often miscarry and stumble on the way'.[46] A poet commented on the same scene:

> Huzza! She rises, – slow – she stops –
> Grows faint, despairs, her wings she drops.
> Then to the earth she sudden came,
> And thus address'd them to their shame,
> 'Your Cause wants virtue more refin'd
> Before admittance you can find
> To higher regions; federal fire
> Can but consume, nor raise us higher.
> Now be convinc'd, by what you see,
> That men were made for liberty'.
> Then sudden bursting to a flame,
> It quickly did consume the same;
> So good by, now, to Adams John,
> And Louis Tousaint's Air Balloon.[47]

Adams, perhaps, is the 'self-made mushroom patriot' referenced by the first commentator. In any case, the failure of the balloon was linked directly with the current administration. In all these instances, balloons offered a negative image. Thus, while balloonists easily promoted their craft as potentially useful, some authors found it much easier to focus on the failures of aeronautics when using balloons as a political metaphor.

Although balloons appeared more frequently in France, Great Britain and the United States, other nations also used aeronautics to promote or critique the state. An attempt in 1788 to forge unity in Poland between the king, Stanislaus Augustus, and those who opposed him included the launching of a hot-air balloon in Warsaw by Blanchard. Blanchard also demonstrated a parachute for the king by dropping a dog, attached to a parachute, from a third-floor window.[48] The Ottomans hired two Englishmen to go to Constantinople and launch a balloon there. The two men – a Mr Baily and Mr Devignes, an artist and member of the Mathematical Society of London respectively – performed a small launch for the Grand Signor who then 'ordered them to make a public experiment on a larger scale'. The British ambassador, Lord Elgin, and all the other foreign ambassadors arrived to see the globe, made of 'red silk, ornamented with crescents and stars in different coloured foils', ascend. It stayed aloft for twenty hours, eventually crossing the Sea of Marmara. The Grand Signor rewarded Baily and Devignes with money and furs.[49] A description of the launch made a point of noting 'the ignorance of the Turks was very apparent on this occasion' and when a 'breeze sprung up which caused it [the balloon] to incline towards the Grand Signior, and afterwards on the recoil towards Captain Pacha – this they took to be an intended salute to those personages'. This account also claimed the Turks were jealous of the ballooning skills of the British aeronauts and that the state 'contrived to introduce two Turks into the ground half an hour before the ascent, in order to persuade the people they had a hand in it'.[50]

Napoleon took a company of balloonists with him on his venture to Egypt. In order to 'awe the people' he had the aeronauts send up a balloon in Cairo.[51] One anonymous source even claims that Napoleon himself went up in the balloon while in Egypt 'to prove his courage and his bravery as well as his knowledge in the sciences and the arts'.[52] In Aranjuez, Spain, a commentator wrote that political news was lacking because 'the public attention has been engaged in the sending up a balloon…under the conduct and direction of M. Baretu, who was embarked therein'. After the launch, however, the author noted Baretu 'was seized with a giddiness in his head, fell to the ground, and broke one thigh and two ribs'.[53] The commentator followed this immediately with the assertion that the new Spanish minister, Count d'Aranda, seemed to have little influence. The balloon accident, sandwiched between a claim of no news and a critique of the new minister, may be coincidental, but it nonetheless manages to conflate the weak minister with the failed launch.

Military Ballooning

The appropriation of aeronautics to the military had enormous potential. Although many commentators favoured the use of balloons in this capacity, the idea frightened some critics.[54] Horace Walpole exclaimed 'I hope these new mechanic meteors will prove only playthings for the learned and the idle, and not be converted into new engines of destruction for the human race, as is so often the case of refinements or discoveries in science'.[55] Walpole's fears, however, were realized. While the state frequently appropriated aerostatics for festivals, and while the public just as often used balloons as a means to critique the state, some savants and politicians worked together to try and fulfil the potential utility of this new invention and find a way to apply balloons to the military service of the state. Thus, although the French state first tried to encourage the application of aeronautics to commerce and, as such, invested 40,000 livres in the Montgolfiers to encourage work in this area, during the French Revolution the state worked together with savants to form the first military balloon company.[56] Catherine the Great of Russia also worried about the possibility of balloon armies against which 'not a single fortress could withstand'.[57] Benjamin Franklin wrote that 'Five thousand Balloons, capable of raising two men each, could not cost more than Five Ships of the Line; and where is the Prince who can afford to cover his Country with Troops for its Defence, as that Ten Thousand Men descending from the Clouds might not in many places do an infinite deal of mischief?'[58]

Commentators proposed military applications almost immediately after the appearance of aeronautics. A letter addressed to the *Affiches de Province* by Giroud de Villette, dated 20 October 1783, claimed that the army would discover 'real utility' of the Montgolfier brothers' invention. He suggested generals use globes to discover the position of the enemy, their manoeuvres, where they were going, and how they were arranged.[59] In November 1783, an essayist writing in the *Courrier d'Avignon* concurred arguing 'the utility, the grand utility' of balloons is 'for throwing light on the march [and] the approach of the enemy army, what occurs in besieged villages, and the topography of a country, etc'.[60] A painting, created by a Mr Burke in London, depicted a balloon invasion. Titled *The Alarm*, it was a 'representation of a hundred thousand French Republicans, who after arriving in London in a Balloon, are ravishing our wives and daughters, poisoning the River Thames, and making daggers to fly about in the air, like snow in a windy day'.[61]

Some authors who expressed concern about the use of balloons in the military often focused on the idea of an aerostatic invasion. During the French Revolution, an actual proposal by a Frenchman named Thilorier, described as a physicist, caused considerable alarm when it appeared in November 1797. Thilorier offered to construct a vast globe 'to raise and transport [an army] to the heart of England'.[62] This aeronautical armada led to quite a furore among British and American newspapers in early 1798.[63] Some contemporaries assumed that Blanchard would be chosen to lead this new floating flotilla although others thought Garnerin the likelier candidate; some, of course, just thought it was nonsense.[64]

Blanchard's name had been mentioned in connection with a 'fleet of balloons' as early as 1793 and was still being bandied about in 1799 when the *City Gazette*, based in Charleston, South Carolina, claimed he was 'at present engaged in forming a squadron of balloons, with which he has undertaken to make an aerial excursion to America'.[65] Thilorier's plan was still a topic of discussion in 1803 when a sonnet was written on the subject called 'On M. Thilorier's plan of invasion by means of Air Balloons, carrying 3000 armed men each!!' The poet ended the sonnet with the suggestion that the British 'Place your best marksmen on the cliffs of Dover / To wing the wildfowl as they're flying over'.[66] In 1804 a rumour spread that Napoleon might use a balloon to secretly land in England so that he might be there when his army, arriving more conventionally on ships, landed.[67]

In 1808, an anonymous American commentator also fretted, in a somewhat joking manner, about the threat of aeronautic invasion from France. The author wrote that 'a French army was on the peake of the Allegany Mountains, conveyed in Balloons, and were descending fast to be joined by "dogs, negroes, and democrats"' who apparently wanted to overthrow the newly established government.[68] A report from Moscow in 1812 suggested that one of the Russian tactics against the invading French would be to throw incendiary devices down on the French army from balloons; instead, the anonymous author suggested, the governor of Moscow, Count Fyodor Rostopchin, used the devices to burn Moscow.[69]

Whatever the imagined use of balloons for the military might have been, the French did create an aerostatic army during the period of the Revolution. Reports made by the Committee on Public Safety argued for the viability of such a force and the response was to create an Aerostatic Development Centre in 1793 followed by a Company of Airmen in 1794.[70] This company was composed of forty men and three officers. Jean-Marie Coutelle commanded the Company while Nicolas-Jacques Conté acted as his second-in-command. Neither of them had much experience with ballooning prior to this time, but they were both keen amateurs. Coutelle had followed Jacques-Alexandre-César Charles' course in experimental physics and Conté was an autodidact and inventor.[71] The company did not see much action overall, but they did play a key role in the Battle of Fleurus in 1794, barely a month after the company had been formed. In that confrontation, Coutelle oversaw the construction and launch of a balloon called *L'Entrepreneur*, which provided information on Austrian movements and the composition and placement of their troops. Enthusiasm was so great for this new role balloons might play that the French created an Ecole des Aérostats in Meudon, directed by Conté, where sixty cadets were to be trained as future 'balloonist officers'.[72]

It is not clear if the balloon regiment played a significant role in the ensuing French victory at Fleurus, but balloons certainly received considerable credit for the success. One commentator wrote 'The French, with their usual ingenuity, have adopted a new mode of reconnoitering our situation – a mode that there is no possibility of preventing'.[73] An illustration of the battle shows the tethered balloon hovering over the action where someone in the gondola could clearly see behind enemy lines (see Figure 4.1). One author claimed 'what rendered this event more extraordinary was, that the French directed all their operations according to the information they received by means of Balloons suspended in the air'.[74] This author's confidence that balloons dominated the field of battle is unwarranted; however, it is significant in his emphasis on the possible influence people believed balloons might have in that arena. The Baron de Selle de Beauchamp, writing in his memoirs, suggested that if nothing else, the balloon had a great psychological effect on their opponents. He claimed 'the effect on morale in the Austrian camp caused by such a new spectacle was immense'. The enemy soldiers, he said, believed the French to be sorcerers.[75]

As mentioned earlier, Napoleon took a balloon company with him to Egypt although they did not see military action there. Instead, he had the company launch a balloon as part of a ploy to impress the native population with French engineering skills. This tactic, however, did not seem to work particularly well as one observer noted the balloon 'was launched in the middle of the absolute indifference' of the Egyptians. They were more impressed when the balloon came falling down and caught fire; this showed promise to the locals who thought the balloon could be used to set fire to cities under siege.[76] When Napoleon fled Egypt to return to Paris and check on his wife, whom he had heard was having an affair in his absence, his troops along with the balloon company remained behind. Ultimately, Coutelle and Conté spent three years in Egypt before they managed to return to France. Upon their arrival home, however, they discovered that the Balloon Company had been disbanded during their absence.[77]

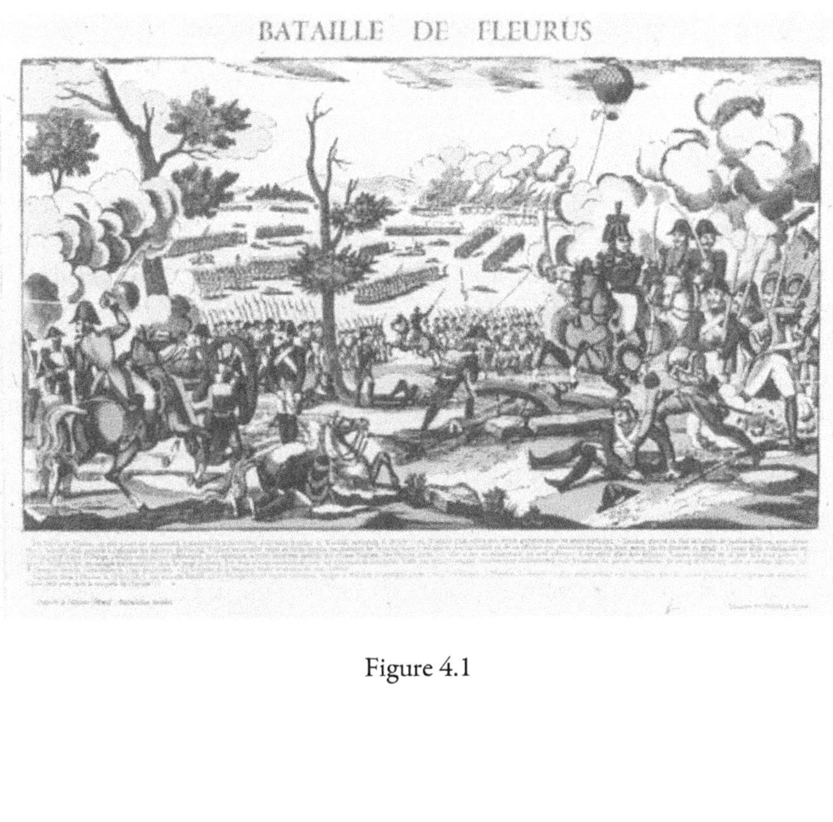

Figure 4.1

Controlling Launches

Although the French state was willing to use balloons themselves, there was also an interest on their part to limit access to ballooning. Of course, the desire to control admission to launches came from many different sources, not least of which was the aeronauts themselves. Certainly it is true that some of the efforts at conducting scientific experiments from balloons were frustrated by the crowds of people who flocked to see every launch. Occasionally, launch times would be purposefully concealed so as to prevent the uninvited from showing up, although this more often occurred as a way to control who did show up and, probably, ensure that they had purchased tickets for the privilege. The famous flight from the Château de la Muette in November 1783, for example, was not advertised openly. Etienne Montgolfier, who was in charge of this launch, did not desire any public announcements which, in the opinion of Faujas de Saint Fond, were 'pompous' and 'humiliating to the sciences'.[78] Nonetheless, not every balloonist wishing to perform experiments could prevent a crowd from gathering. On one occasion, the audience actively thwarted an experiment on the nature of thunder and the transmission of sound when their shouts drowned out the noise made by the balloon exploding.[79] Blanchard had all of his meteorological equipment broken when a young cadet from the Ecole Militaire tried to force his way into the gondola of Blanchard's balloon and join the flight. During the French Revolution one satirist attacked the up-and-coming Napoleon Bonaparte by claiming, falsely, the cadet in this instance was the young Napoleon.[80]

With such examples, it comes as no surprise that some savants tried to keep their launches a secret. Zambeccari, for example, tried to limit the number of people at his launches. He would not announce the day or time of the launch but instead printed that information on the tickets; in addition, he pleaded with ticket holders to 'forbear communicating' their knowledge to the public in order to 'prevent the inconvenience of too numerous an assemblage of spectators'. However, rather than an effort to focus on science, this seemed equally an attempt to prevent a riot in case of failure and to ensure that all those who saw the launch had paid for the privilege.[81] For the first manned launch in Paris, with Pilâtre de Rozier and d'Arlandes doing the honors, the *Journal de Paris* announced that the experiment was strictly scientific and would be of interest only to savants. This had little effect in limiting the size of the audience that appeared to watch the launch. It also was patently false since the entertainment value of science has long since been established and was a constant theme in the pages of the *Journal de Paris* itself.[82] In fact, a significant lack of direct patronage on the part of the state left aeronauts dependent on the general public and also free to ignore doing much in the way of scientific experiments during their flights. Instead, aeronauts found themselves able to conduct launches that favoured the entertaining over the scientific.[83]

Perhaps with the memory of his run-in with the French cadet in mind, Jean-Pierre Blanchard performed his 1793 launch in Philadelphia, the first manned launch in North America, from inside the walls of the Walnut Street Prison. This venue afforded him multiple protections, not least of which was protection against the strong winds that might have damaged the balloon during the process of inflation. More important, however, constructing, inflating and launching

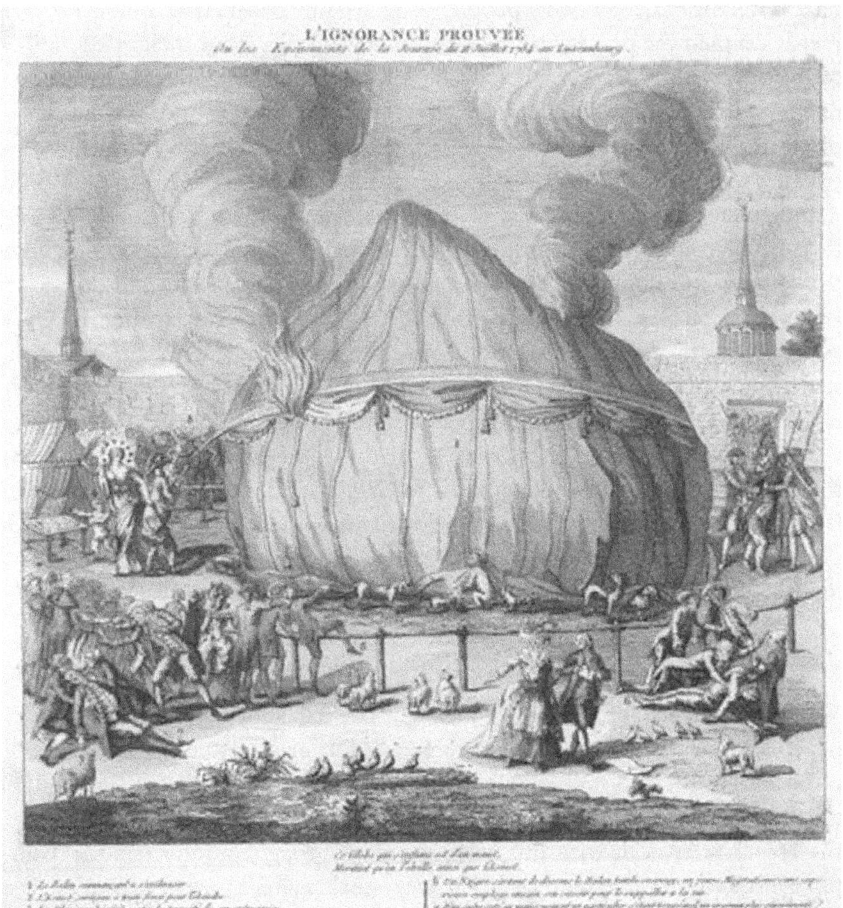

Figure 4.2

his balloon from inside the prison kept vandals at bay and allowed Blanchard to control access to the moment of the launch. As Blanchard's livelihood depended entirely on ticket sales, this level of control was absolutely essential.[84]

Charles and the Robert brothers had to call in guards, both on foot and horseback, to maintain order for their experiments. For their unmanned launch from the Champ de Mars Charles hired troops to cordon off the area. Only ticket holders were allowed to get inside the perimeter and, therefore, get close to the balloon.[85] For his manned launch, Charles needed troops stationed in and around the Place des Victoires near his workshop and one of the locations where tickets were for sale. So many people rushed to buy tickets that Simon-Prosper Hardy claimed the police were needed 'to prevent the aeromaniacs from suffocating each other'.[86] As the launch itself took place in the Tuileries gardens, it was relatively easy to control access and guards were set up at four of the entrances to ensure that only those with tickets were admitted. Lunardi had wanted to release a balloon from the courtyard of the Chelsea Hospital, but the Hospital withdrew its permission because of a fear of rioting. Lunardi instead set his balloon afloat from this training field of the Artillery Company in Moorfields with the soldiers also doubling as guards, along with the city militia and others, to keep the peace. Not all balloonists had the advantage of a walled garden, military company or prison yard. Thus, not surprisingly, a number of aeronauts employed armed guards and built fences to try and establish a perimeter around their launches. Figure 4.2, for example, clearly indicates a railing used to try and keep the audience away from the balloon; in this case, the barrier did not work and the audience destroyed the balloon. Antoine Rivarol described Charles's launch from the Champ de Mars as being cordoned off by a circle of guards. The police even had to escort Charles's balloon from the shop where it was constructed to the launch site.[87] This did not always work as well as they would have liked.

State Legislation on Ballooning

While balloonists approached the control of aeronautics in a variety of ways depending on their scientific or commercial leanings, governments in general responded to the new invention of ballooning by attempting to regulate it. This did not happen everywhere and occurred for a variety of stated reasons. Initially, the French state dealt with this new invention through an informational pamphlet printed on 3 September 1783, very quickly after the first launches. The goal of this short pamphlet was to prevent the terror the people might feel on seeing a balloon in the sky for the first time. The state felt it necessary to warn people because of the reaction of the citizens of the village of Gonesse who had attacked and destroyed the hydrogen balloon released by Charles and the Robert brothers from the Champ de Mars on 27 August 1783.[88]

In France, the state tried to limit who could conduct launches as well as where and when they took place. The lieutenant-general of police in Paris issued an ordinance on 23 April 1784 which said that no one could build or launch balloons or any other aerostatic machine because of the possibility of fire. Hot-air balloons, the ordinance claimed, had landed in the Tuileries Gardens, the Quai des Théatins, as well as in a variety of other places, all replete with combustible matter. As hot-air balloons often carried with them a small fire to keep the air inside the balloon hot, the potential for disaster was very real. Thus, the police of Paris, in order to prevent such accidents, forbade the construction and use of balloons both by experienced aeronauts and those without knowledge or experience. The ordinance went on to say that even people of rank and status were to refrain from tinkering with balloons. If you wanted to set up a launch, you were supposed to ask for and obtain permission in advance by submitting a request naming the place, day and hour of your experiment although the ordinance suggested that only people with experience and knowledge would get such authorization. Anyone launching a balloon without doing this would be fined 500 livres, a considerable sum for most people at that time.[89] The *Gentleman's Magazine* commented on this French law favourably, noting that 'the necessity of this restriction on the madness of fashion in this instance will appear from an accident which happened by the fall of an air balloon, to which burning lamps had been appended, in a gardener's garden near Isleworth. It was found burnt to ashes, and had communicated its flames to the gooseberry bushes round it'.[90]

The police in Angers, France concurred with the Paris police and issued their own version of the Paris ordinance on 19 May 1784. This law repeated much of the same information as found in the Paris version; all people, regardless of status or rank, were forbidden to make or launch a balloon without first obtaining permission. Those found guilty of breaking this law would be fined 500 livres. The authors of the ordinance in Angers, however, went on to suggest that the Paris police had found it necessary to halt the abuse of a discovery which, while it may be useful in itself, could also be extremely dangerous in the wrong hands.[91] Thus, while the people of Angers were being protected from the danger of fire and accident, the officials there were equally clear in their desire to limit the use of balloons to those with a proper background. As in Paris, the authorities in Angers wanted to make sure that not just anyone played around with balloons; only those deemed acceptable by the state would be given permission.[92]

The state authorities apparently alerted all the local Intendants in France and suggested they adopt a similar position. Count Vergennes, the Intendant of Caen in Normandy, was told that since balloons 'may cause very harmful accidents' the King intended that the Police of each region should 'prohibit the launch of aerostatic machines' that had a fire on board unless the balloonists first obtained permission from the government.[93] Some local officials expressed surprise at this edict. In a few of the areas around Caen, for example, the officials claimed that their citizens were not occupied with aeronautics and that there were no plans to launch 'even the smallest balloon' in that region.[94]

Nonetheless, most areas did respond with laws of their own. In Caen, for example, the local lieutenant general of police published a slightly modified version of the Paris ordinance. He wrote that 'the enthusiasm that surrounds this novelty' has caused the public to 'close its eyes to the accidents that are inseparable' from balloons 'in the hands of non-experts or the overly daring'.[95] The law in Caen was tested less than a month later when Eustache Duval wrote to request permission to launch a hot-air balloon. The request went all the way to the Intendant, Vergennes, who received warning that Duval was not a very good mechanic and did not have great knowledge about balloons and, therefore, that the request should be turned down; subsequently, Vergennes denied the request. Local communities in Normandy were still issuing ordinances against ballooning in 1785; in that year the town of Bayeux published a law noting that balloons, moving at whatever direction the wind took them, could hit houses and cause accidents. Therefore, ascensions without permission were once again forbidden.[96]

Other parts of France were equally vigilant in their attempts to manage aeronautics. Strasbourg instituted an ordinance against ballooning just as two savants Degabriel and Pierre, along with one of their workers named Hersant, were preparing to launch a manned flight from that city. Alarmed that their work would go to waste, Degabriel and Pierre petitioned the local magistrate. They managed to get a number of other people to attest to the 'excellent construction of the balloon and of the minute precautions' they were taking in case of any accident caused by fire. As a result, the local government granted them an exception and allowed their launch which took place on 23 June 1784.[97]

The authorities in nearby territories also regulated balloons. In what is now Belgium for example, a number of towns issued edicts forbidding launches without the permission of the local magistrates. Ypres did this in 1784 and Malines in 1786. Apparently, the town of Namur actually had a fire caused by a hot-air balloon in 1785 which led directly to their ordinance against aeronautics. All of these local examples eventually led the national government, at the request of the Privy Council, to issue an edict banning launches that had not been given prior permission. This law had first been considered by the national government in 1784, but was not passed until 26 May 1786. As stated in the ordinance, it was the fear of fire that most inspired the state to pass this law.[98]

The *Monthly Review* reported that Frederick II the Great of Prussia had prohibited 'all aerostatic experiments in his dominions'. The newspaper went on to suggest, somewhat tongue-in-cheek, that since Austria and Russia dominated on land, England at sea, and now the French in the air, it was left for Prussia to try and dominate fire.[99] Somewhat more seriously, the *Essex Journal* suggested that Frederick argued it was 'useless for his subjects to partake of this glory of this discovery, since one cannot aspire to it, but at the risque [*sic*] of his life'.[100] This ban was continued by Frederick's successor, Frederick William along with the Holy Roman Emperor Joseph. An American newspaper reported that Joseph, in talking about Blanchard, was impressed by the success of ballooning, but that it would not be until Blanchard had 'found out a method by which you [Blanchard] can render those travels ... useful, that you can afford me any pleasure in coming to Vienna to instruct me on the subject'.[101] The *American Recorder* also used these bans to poke fun at the Prussian monarch along with his Austrian counterpart. '*Balloon flying* is forbid in the emperour's [*sic*] dominions, and in Prussia' the paper wrote; 'the wise Joseph, and the cautious William, have forbid Blanchard to hunt any of their *wild geese*; or set their subjects an example so much at defiance with German *gravity*'.[102] A Russian newspaper claimed that Frederick was worried about balloons 'in the hands of inept people' becoming dangerous and causing fires. Thus, no launches would be allowed without the express permission of the state.

Similarly, in Milan launches were banned after several rogue balloons landed near locations including one that crashed near a gunpowder magazine. 'Only the most skillful persons alone may receive permission to make experiments on occasion', the law stated, 'but with precautions required for the observance of public security'.[103] Similar bans were issued by Wilhelm IX, landgrave of Hesse in 1785 and by the Swiss Canton of Solothurn in 1784.[104] A decree published in 1786 in Brussels also forbade hot-air balloons.[105] Reportedly, the lord mayor of Dublin had prohibited balloons in that city because of 'the repeated mischiefs they had occasioned'.[106] The magistrates in Newcastle even went so far as to offer a reward, of fifty pounds, 'to any person who will give information of any one who may hereafter be guilty of setting off balloons made of paper, with some burning matter affixed to them' something they deemed to be 'a practice of dangerous tendency'.[107] A balloon in Gloucester came crashing down, all ablaze, landing just a few yards from a stable door, where some straw was strewn which could have set the whole building on fire. This incident, along with other accidents that had happened in Kent, encouraged the magistrates there to prohibit balloons.[108]

Catherine the Great also banned balloons in her realm. Although there were a few unmanned launches in the early period of ballooning, from 1783 to 1785, in 1786 Catherine 'personally vetoed a proposed visit' by Blanchard who by that time was one of the most famous balloonists in Europe thanks to his feat of having crossed the English Channel from Dover to Calais. Royal disapproval meant that the first manned flight in Russia did not take place until 1803 when

Garnerin and his family of balloonists travelled to St Petersburg and went up in a balloon in the presence of Alexander I.[109]

The fear of fire was very real, even if the number of balloon-related fires was fairly small. The first public launch by the Montgolfier brothers had ended in a fire. One of the problems stemmed from the inability to steer balloons; thus, they might land anywhere and start a fire. The first aeronauts, Pilâtre de Rozier and the Marquis d'Arlandes, recognized this danger but had little to offer to assuage the fears of the state. Their somewhat uninspired solution was to bring with them a pail of water and two sponges although this was mostly to keep the balloon itself from catching fire.[110]

As a side note, while the state was almost uniform in its claim to want to protect their cities from fire, they rarely expressed concern over the lives of the balloonists themselves. The risks taken by aeronauts were, apparently, their own concern. Only Louis XVI had suggested, for the first manned flight, sending up a condemned prisoner rather than a savant in case of accident. This idea was quickly set aside as Pilâtre de Rozier had no intention of letting a criminal take the glory as being the first man to sail in an air balloon. Had Pilâtre de Rozier foreseen his own fate just two years later, as the first man since Icarus to die as a result of falling from the sky, he might have been more cautious.

Balloons and Crime

In addition to fires, and sometimes the lives of balloonists, states had other concerns about the practice of aeronautics within their realms. Aeronauts drew enormous crowds of spectators and 'established a direct and unmediated relationship' with 'enormous multitudes of people'.[111] But unlike other large gatherings, such as religious or political processions, scientific crowds were somewhat unpredictable, as was the science they gathered to watch. Simon-Nicolas-Henri Linguet suggested that balloons seemed to him 'an instrument appropriate for facilitating larceny and abductions'.[112] Bachaumont reported that the Paris police were worried about the large number of 'malcontents' who used balloons as 'a pretext for gathering together in large numbers'. As a result, the police requested that aeronauts not post any notices about forthcoming launches and, in particular, that they not let the general public know the day of launches since the police force was not large enough to ensure the 'public's security'.[113] Pilâtre de Rozier got around this order by holding one of his early tethered launches on a Sunday (so that wage-earners could attend without losing work) and by sending letters to all the subscribers of his *Musée de Monsieur* giving them the particulars of the launch. This flight, which was scheduled to leave the gardens surrounding the Réveillon factory at 4:00 pm, drew a crowd of around two thousand official guests, that is, those let in by the police because they had tickets. But a large police guard was also required to prevent admission to those without tickets.

In some cases, the problems arising from crowds around launches were concerned only peripherally with ballooning itself. In particular, it turns out that large groups of people all looking up provided exactly the kind of situation designed to benefit thieves; the 'general cry of *pick-pockets*!' might not have been an uncommon occurrence on such occasions.[114] The *Morning Post* reported in 1784 that at Mr Astley's proposed launch at St George's Fields an enormous 'concourse of people' were assembled together, perhaps more than had ever been assembled on any other occasion and that the crowd included 'spectators of all descriptions, from the Peer to the pickpocket'.[115] The *Gentleman's Magazine* suggested that all the spectators at Astley's launch would remember it since 'a more ample harvest for the pickpockets never was presented' adding that 'some noblemen and gentlemen lost their watches and many their purses'.[116] Henry Laurens also commented on the propensity of pickpockets to frequent balloon experiments; in a letter to Benjamin Franklin he wrote that he decided not to attend a launch and thus 'saved [his] watch by staying at home'.[117] At a launch by Lunardi the crowd caught a pickpocket and ducked him in a nearby pond.[118]

Many 'people of fashion' attended a ascension by Blanchard on 9 May 1785 according to the *London Times*, but the crowd also included several pickpockets. The thieves were 'remarkably successful, many gentlemen and ladies having been robbed; among them was Mr. Edmund Burke, who lost his purse and watch'.[119] The newspaper later suggested that the government place a heavy tax on balloons which appear to have no other use than to fill the pockets of aeronauts at the expense of the public. This switch, of naming the balloonists as the thieves, was echoed by the *Publick Advertiser* which suggested that foreign aeronauts, in this case Blanchard, were duping the British and hoped to make a 'dive into [their] pockets'. The *General Advertiser* made the same association between balloonists and pickpockets commenting that aeronauts had found that 'the best implement for picking pockets' is 'cotton dipped in spirits of wine', (which was used to keep the air hot in some balloons).[120]

The British public also had to worry about other kinds of theft; Richard Crosbie found himself forced to reissue tickets for his January 1785 launch when it was discovered that forgers had been selling faked tickets and passes.[121] In other instances, tickets were sold for performances that never happened. This was the case in New York where several hundred people bought tickets, at fifty cents each, for a Lunardi launch that never happened.[122] Thieves were widely reported to operate at French launches as well.[123] The French system, based on subscriptions, could lead to additional problems. In December 1783 the *Courrier d'Avignon* reported that a 'charlatan de physique' had raised 1,500 livres for his experiment; but when he saw that it wasn't going to work, he disappeared.

This echoes the kind of charlatanry that occurred in other instances such as the man in Paris who raised a subscription to fund his attempt to walk across the Seine using 'elastic shoes'.[124]

Balloons and Riots

While fires and theft certainly loomed as potential problems associated with launches, the most significant threat to states came in the form of riots. Large numbers of people congregated at ascensions. Many of them had travelled great distances; many of them had paid for the privilege of seeing the balloon up close. However, aeronauts had frequent difficulties in meeting the expectations of their audience. As one commentator noted, 'a person who offers to go up in a balloon for the entertainment of the public, has a double chance of meeting with death'. If the launch succeeds, he must face the dangers of the voyage; if it fails, 'he must fly to avoid being knocked on the head by the mob'.[125] Hot-air balloons were notoriously tricky to inflate depending, for example, on wind and humidity. Hydrogen balloons required considerable expertise to manufacture the necessary quantities of the gas. In both cases, poorly constructed balloons allowed hot-air and hydrogen to leak out before the balloon was fully inflated causing further delays. In all, the people who attended launches were often lucky if the experiment occurred on the same day, much less the same hour, as had been announced. This tended to make crowds somewhat unruly and, not unexpectedly, riots occasionally erupted. Franklin, for example, recounted a story he had heard about a failed launch in Bordeaux. He noted 'it is a serious thing to draw out from their Affairs all the Inhabitants of a great City & its Environs, and a Disappointment makes them angry'. He described how someone had 'pretended to send up a Balloon & had received Money of many People', but that when the balloon failed to launch 'the Populace were so exasperated that they pull'd down his House, and had like to have kill'd him'.[126]

One of the biggest riots happened early on in the history of public launches and thus established rioting as a realistic and viable response to failure. Crowds now had a model to follow if launches went poorly and aeronauts had a tangible example to instill fear in them lest their efforts not meet up to the expectations of their audience. This riot arose from the misguided efforts of the self-styled Chevalier de Moret who announced a launch from a garden in Chelsea on 11 of August 1784. Moret, who claimed acquaintanceship with the Montgolfier brothers, was trying to out-do Lunardi, who had announced his flight for the next day. 400 people paid the subscription price to see Moret launch his balloon while an estimated 50,000 additional people congregated in nearby streets and fields. The launch, planned for 1:00 pm did not take place until 4:00 pm due to problems with the inflation process. The *Morning Herald and Daily Adver-*

tiser reported that Moret's nerve gave out just as he stepped into the gallery of the balloon; but that did not really matter since 'nearly at the same moment, as misfortune would have it, the rope which held the Balloon gave way, and, to the surprise of every beholder, it found more attractions to the earth than the air, and immediately came to the ground'. The audience did not react positively to this outcome:

> Disappointment was visible in every countenance, and that part of the audience who are not remarkable for their philosophy, resolved to have satisfaction of the deceiver, and for that purpose pulled down the gates, burst into the garden, and, with the rapacity of hounds tore their prey into a thousand pieces; their rage, however, was by no means satiated by the destruction of the Balloon; the fire which had been kindled for the purpose of procuring inflammable air, was made an instrument of their revenge, and in a few moments, the seats, pales, and every thing they could lay hold of that was consumable, were committed to the flames.[127]

While the crowd took out its dissatisfaction on the balloon, and marched 'off with pieces of the Balloon, as trophies of their victory', Moret seems to have gotten away with his body, if not his reputation, intact. Moret took advantage of the melee to make good his escape from the area. This example shows the audience behaving with a sort of moral economy of the scientific crowd. In the absence of a planned scientific experiment, they took it upon themselves to act out the part of the aeronauts and 'launch' their own balloon complete with the moment of ascension and the final crash.[128]

Lunardi had few good things to say about Moret. His own flight had been scheduled for the next day from the gardens of Chelsea hospital. But the crowd had spread such 'desolation and terror through the whole district' that, unsurprisingly, the governor of the hospital was unwilling to let his grounds be used, and potentially abused, by a launch. Lunardi quickly found a new site, the Artillery Company training grounds, which served both to protect him and his balloon and keep the crowd pacified. Lunardi also had to distance himself from Moret and claim his own flight was much more scientific and, thus, potentially successful. The *Political Magazine* wrote that 'notwithstanding the failure of Mons. Moret, another aeriel navigator, Mons. Lunardi promises soon to indulge the public curiosity'.[129] The same journal also noted that when Lunardi finally performed his launch, he took off in the balloon when it was only three-quarters full. He felt 'intimidated by the reports of the fury of an English mob' and did not want to risk waiting any longer.[130] In his account of the day, Lunardi wrote that he had 'reason to be anxious not to disappoint' the multitude 'every one of which has been wedged in a painful situation the whole morning. '[131] Luckily for Lunardi, the Prince of Wales attended and spent much of his time near the balloon even though, according to Lunardi, the Prince's retinue expressed 'anxieties for his safety'. Nonetheless, Lunardi felt, the Prince's presence removed 'the suspicion of deceipt' and restrained 'the impetuosity of the people'.[132] Lunardi continued taking precautions during his series of launches in Scotland two years later. For one such event he 'requested the attendance of the military and ordered the gun to be fired and the flag hoisted. The fear of disappointing the *public curiosity*, which is now highly raised, has outweighed every other consideration'. Lunardi added that even though the local Lord Provost himself promised to keep the crowd under control, he 'declined the offer, and determined at all events to ascend'.[133] Although Lunardi received support from local officials, he ultimately had to flee Britain entirely after a young man named Ralph Heron died during a demonstration.[134]

While Moret was one of the first aeronauts whose failure to launch led to a riot, he was certainly not the only one. A Birmingham barber named Harper, for example, purchased a balloon from James Sadler with the intention of conducting a launch 11:00 am on Wednesday, 29 December 1784. The day prior, in anticipation of the launch, 'strangers of every denomination, in carriages, upon horses, and on foot, crowded into [Birmingham] from every part of the Country'. By Wednesday an estimated 60,000 people were waiting, some of whom had travelled as far as forty or fifty miles. At 9:00 am things were not looking well for Harper who was having trouble filling the balloon. Initially, when 'at length the hour appointed for the balloon's ascending having elapsed' it was reported that 'the populace discovered no kind of impatience, being assured that it would certainly go up in the afternoon'. However, the mood of the crowd quickly changed when Harper continually failed to inflate the balloon. 'This disaster being inculcated amongst the multitude, they instantly became clamorous, and soon proceeded to outrage, throwing sticks, stones, dead dogs and cats, &c. over the scaffolding erected for the accommodations of such persons as had paid for seeing the balloon filled'. Things got worse and 'peace officers' were called in; they attempted to seize the most active members of the mob which caused only to escalate the violence. 'One of the constables received a dangerous contusion' on his head and one of the mob had his skull fractured and later died. Things reached such a head that local authorities had to 'read the Riot act'; only then did the crowd disperse.[135]

While not all failed launches led to the throwing of dead dogs and cats, riots of various levels of intensity were not uncommon. The Abbé Miollan and his partner Janinet had their balloon torn to shreds and they were run out of town after their failed launch from the Jardin de Luxembourg in Paris during the summer of 1784 (see Figure 4.2). They had established a subscription in advance and had advertised broadly for many months prior to the event. Although the launch attempt did not take place until August, a prospectus announcing the ascension appeared as early as February.[136] Miollan, described as a 'professor of physics', and Janinet, an 'artist and physicist', sold tickets for six livres each. This was somewhat on the high end for balloon tickets, but it came with a three-day schedule during which the aeronauts would lecture on the science of aeronautics and allow people in to watch the construction and inflation of the globe.[137] Miollan had offered some public lecture courses in 1783 and had worked with Jean-Paul Marat in his public lecture courses.[138] However, he and Janinet were unable to succeed in their launch. They were using a hot-air balloon; but it was a hot day in August and the balloon would not inflate.[139] The crowd quickly turned on the would-be aeronauts, destroyed the balloon and forced them to flee for their lives. Anna Francesca Cradock attended this ill-fated launch attempt by Miollan and Janinet and describes the audience as 'a compact mob' even before the launch failed; as evidence of this she notes the difficulty she had in getting to her seat. Once the experiment was called off she and her companions settled into a nearby café where they had bread, wine, a salad, and a very expensive chicken which they enjoyed in good humour. However, the rest of the audience behaved in a different fashion, forcing the gates, burning the chairs and tearing the balloon to pieces. There was even some worry people might attack the palace.[140]

Miollan and Janinet became the object of many jokes and satires in the days and months following their failure. Many puns were made about Miollan's name, associated it with a cat (mewing), as well as Janinet's name (an ass or âne). Nikolai Karamzin, Russian traveller visiting Paris, recounts a story he heard about a man standing on the parterre at the opera when a tall abbé came in and stood in front of him. To get rid of the troublesome man one of his friends shouted 'There's Abbé Miolan, who deceived the public'. Although the abbé protested his innocence, he was nonetheless forced to flee.[141] It did not help matters that Miollan and Janinet came across as pompous.[142] This made their failure even more delightful in the hands of their critics.

The would-be aeronauts Dr Grassi and his nephew Perier of Bordeaux also had trouble with their ascension from the Royal Garden in May 1784. 4,000 subscribers turned up along with an immense mob of the curious who gathered just outside the launch site. Strong winds, however, worked against Grassi and Perier. After the time for the launch came and went, a part of the mob outside the gates stormed the barricades; ultimately, two hundred chairs were burnt and several of the individuals guarding the launch site were beaten. Grassi and Perier barely escaped, finding refuge in a café in the garden.[143]

Blanchard had his share of problems as well. An English audience had expressed its disapproval after his failed descent in a parachute while fiddling. After this disaster, Blanchard left quickly, leaving his audience to take its 'revenge on the machinery and railing of the enclosure, both of which were in a short time demolished'.[144] Blanchard also had trouble in Vienna in 1791 when his planned ascension failed; the crowd expressed some anger until they were told that another ascension was planned and that they would 'be admitted without paying anything' and that in the mean time the money collected would be held by the police rather than given to Blanchard. Even with such precautions, Blanchard had to be escorted to the Police Hall with a guard of thirty men.[145]

The *Morning Post* reported an unsuccessful launch from the Netherlands. While the aeronaut had hoped to soar 'above the earth' he 'thought proper to get under it, by jumping into a cellar, to avoid the resentment of the populace'. The balloon, 'on which was beautifully painted an Eagle with a Thunderbolt in his talons, was torn to pieces by the disappointed crowd'.[146] An apothecary's misguided attempt to launch a balloon in Freiburg, Saxony led the spectators to pelt 'him pretty severely with snow-balls and pebble stones'.[147] In New York City, one gentleman 'unable to brook the disappointment [of a failed ascension], discharged his ire on the head of the aeronaut' by hitting him on the head with his cane 'to the no small diversion of the numerous spectators'.[148]

Time did not lessen the response of frustrated audiences. In 1812 Jacob Degen attempted a launch from the Champ de Mars in Paris. After the balloon failed to launch the police had to intervene to prevent the onslaught of total chaos.[149] In 1802 an aeronaut named Mr Barrett tried to launch a balloon from Greenwich before 'an incalculable number of persons'. When he failed to launch it on the first day it was rescheduled for the next, with the same large crowd turning out to watch. When it became clear that 'after a variety of efforts ... the balloon could not be sufficiently filled with gas, or brought to bear, to take up the parties intending to make the aerial excursion', then all that 'could be done to satisfy the multitude was, to fix a cradle, instead of the car, to the balloon' after which 'the ropes were cut, and the balloon ascended'. In this way, Mr Barrett avoided a potential situation.[150] In 1819 the citizens of Philadelphia suffered a riot after Monsieur Michel's balloon failed to take off from the Vauxhall Gardens. People threw stones while a boy, who had climbed on top of a fence, was struck down. The crowd took out their anger on the balloon and accompanying parachute, which were torn to shreds, and also levelled the fence that had surrounded the enclosure. A nearby bar was torn down and the young men present took the opportunity to drink the liquors found there. Eventually, the Pavilion from which the event was to have taken place was burned to the ground. Only three of the rioters were Philadelphians, or so they said, and since another three were young and respectable their names were not published with the others.[151]

Significantly, audiences responded with violence, but generally directed their violence at the balloon or the structures around the balloon, rather than at the aeronauts. In effect, the audience knew best how to respond in such a way that they could take out their frustrations against balloonists for failed launches, while leaving it less likely that the authorities would have to call in police or troops to quell their activities. In other words, the audience applied their existing knowledge of how to riot, drawn from such experiences as grain riots. The audience intuitively knew how best to respond to such events and understood the advantages of avoiding directly harming individuals. This is not to say that some people did not suffer, or even die, during balloon riots; however, such physical punishments were generally circumvented.

In general, of course, observers assumed that the real trouble arose when artisans or peasants attended launches. It was assumed they did not have the patience or understanding their more elite counterparts apparently enjoyed. This stemmed, commentators imagined, largely from their lack of understanding as to what exactly was going on. The *Gentleman's Magazine* reported in 1784 that at one of Lunardi's ascensions the mob was 'on tiptoe' and ready 'to trample the impostor with all his apparatus' while 'the real friends to science trembled for the fate of a worthy man'.[152] Alternately, in 1785 Zambeccari experienced different responses, based on social standing, after his failed attempt to launch a balloon in London. The more elite members of the audience 'were satisfied with Zambeccari's endeavors to please them, and only regretted that his hopes were frustrated'. The rest of the audience, however, was less pleased: 'they tore down the scaffolding, reduced part of the wall to ruins' and would have 'committed more mischief, but for the appearance of the guards'. In the end, 'fourteen of the rioters were taken into custody'.[153] Such arrests were not entirely uncommon. The riot that ensued after Harper's failure in Birmingham resulted in four men locked away 'in the Dungeon for assaults, and damage done to the platform, &c. in the Tennis-Court, and one man for assaulting Mr. Freeth, the constable'. The *Birmingham Gazette* noted that the local magistrates had not punished those arrested since they were unwilling to punish a few 'for that faults that many others were concerned in'. Such leniency, however, was not to be the norm; the newspaper commented that they hoped 'the extreme lenity shewn to the above Men, will prevent similar disturbances, upon the day when the Balloon is to ascend: And have the authority from the Magistrates to inform the public that all persons guilty of the like Offences, will be punished according to the law'.[154] Some anonymous balloonists in Boston, Massachusetts tried to launch a balloon; but when the winds prevented them from getting it off the ground, the lower-class members of the audience 'notwithstanding the exertions of many of the gentlemen present, rushed upon, and destroyed the balloon entirely'.[155] Here again, the moral economy of the scientific crowd demanded a show in spite of the protests of those, more educated, people around them.

When launches failed to take place, the audience, it seems, took it upon themselves to perform their own ritualistic and theatrical performance. This was not so much a dialogue between the aeronauts and their audiences as it was a monologue in which, if the balloonist forgot his or her lines, the audience felt free to pick up where they left off. Instead of a moral economy of the crowd, in which the mob felt that they had the obligation to force a universal morality upon something, such as the sale of grain, here we see a scientific economy of the crowd. In this instance, the audience felt sufficiently enabled by their knowledge of what was supposed to happen that they could model it themselves. The balloon was supposed to be the center of attention through a launch, thus if the launch failed the globe could still provide a focal point through a massive conflagration. Alternately, the audience could mimic the voyage of the balloon by carrying the gondola and balloon through the town, as they did with Tytler's balloon in Edinburgh. Such theatrical performances on the part of the audience replaced the failed entertainments of the aeronauts. The state, of course, frowned on these counter-performances as it once again took the power of science and placed it in the hands of the general population rather than in the hands of the educated elite. Scientific performance was meant to support the power of the state, not the people.[156]

A successful launch with no problems was, occasionally, cause for comment. Thomas Baldwin thought it important to thank the inhabitants of Chester 'for their polite attention on the Day of Ascent, and Preservation of Order during the Inflation: on which the Success of aerial Experiments so much depends, and throu' the Want of which, so many have already failed'.[157] James Sadler, in 1812, wrote:

> It is gratifying to reflect, that notwithstanding the vast concourse of people which thronged not only the Lawn of Belvedere-House, but every Avenue leading to the place of ascent; crowding the Roads, and cloathing [*sic*] the tops of every House from which the slightest view could be had, not a single accident appears to have taken place.[158]

The success of this launch, intended to cross the Irish Sea from Ireland to England, was not necessarily owing to the nature of the people themselves, however, but was thanks to the 'Military for the propriety of their conduct' and 'to the Peace Officers who attended' and 'the vigilance of their attention'.[159] Sadler had constructed a temporary railing to help keep order and used a horse guard to keep the area around the balloon free from 'the pressure of the anxious Spectators'.[160] A police report discussing the 1803 ascension by Zambeccari in Bologna made special mention of the 'admirable order and calm which reigned through all the parish'.[161]

Thus, state edicts seemed to focus on riots, crime and fires as the biggest problems requiring the control of aeronautics. However, the language of the edicts also suggests the state had other concerns about this new invention. The ordinances specifically listed all the types of people forbidden to do launches including commoners and members of the elite, the learned and the unlearned. Surely it galled the state, and their representatives in various national academies, that anyone, of any rank, and with minimal scientific background could conduct a launch. This failed the state's test for useful science in the Old Regime – ballooning failed to reflect the glory of the monarch in the same manner that more elite sciences could because it was too easily duplicated by average citizens.[162]

Ordinances also reflected the state's concerns about the ultimate utility of ballooning. As Emperor Joseph suggested, he would not be interested in balloons until they could be shown as useful, a sentiment shared by many others. Consequently, the state did not offer much patronage to aeronauts since they did not see balloons as a potential servant of the national interest.[163] Laws attempting to regulate balloons did, at least, attempt to keep them out of plebian hands; ballooning, from the state's perspective, was bad because it was popular. But the state did not follow through on this and actually sponsor ballooning through direct patronage or through royal academies.

However, the popularity of ballooning, and its dominance of the scientific public sphere in the period after its invention, did lead the state to make some concessions about aeronautics and use balloons in a variety of public venues. Public opinion most definitely favoured ballooning. Thus, balloons appeared at festivals, coronations, and other public events. But the state ultimately did not offer uniform or consistent support for aeronautics. Ballooning was a bad symbol with which to associate the state, and particularly those states striving for a certain level of absolutism such as France, Prussia, Russia and Austria. Balloonists could not steer their balloons; this was clear from the various ordinances which pointed out the randomness with which balloons might crash land. No monarch wanted unpredictability and a lack of direction associated with their reign.

*****

In sum, a variety of people tried to assert some level of control over the new invention of ballooning. Significantly, few savants claimed the need to control launches for scientific purposes. Instead, aeronauts wanted to control access for financial reasons. Balloons were funded largely through subscriptions and ticket sales; thus, the only way a balloonist could possibly recoup his or her expenses was by making sure that as many people as possible who saw a launch paid for the privilege. This meant, however, that they also needed to find ways to eliminate the ability of non-paying spectators to view too much of the balloon. The state had even more interest in controlling aeronautics. Balloons were potentially dangerous in a variety of ways including as a source of possible fires. Perhaps more perilous, the crowds attending launches were frequently unruly leading to violence and property damage. The state certainly wished to reduce the number of possible riots. In Britain, with the memory of the Gordon riots still strong in people's minds and in all of Europe, where grain riots were common in the 1780s, the concern over possible unrest at launches was all too real.

5 CONSUMING BALLOONS

In the fall of 1785, Count Zambeccari, an Italian nobleman living in London, advertised in the *Morning Post* the imminent launch of his balloon, currently exhibited at the Lyceum, by noting the availability of 'tickets to partake of the peculiar pleasure inseparable from aeriel [*sic*] evolutions.' In addition, Zambeccari pointed out that the best accommodations would be 'under cover, provided with fires, and perfectly comfortable'. Ticket prices ranged from five shillings to half a guinea.[1] This advertisement was one of many that appeared in England and across Europe after the discovery of aeronautics during the summer of 1783. As the historian L. T. C. Rolt has suggested, 'it soon became evident that there was money in ballooning. Not only would people pay to witness an ascent from close quarters, but also there were wealthy enthusiasts prepared to finance a flight and reward the pilot handsomely in return for a place in the car. So there came into being a class of professional aeronauts'.[2] Significantly, the invention of ballooning also coincided with a period of rapid increase in consumerism. Studies of the idea of a rise of consumption have increased dramatically in recent years focusing mostly on the eighteenth century (on the development of consumption or a consumer revolution) and the nineteenth century (regarding mass consumption).[3] For the early modern period, many of these works have in common the idea that a marked change in patterns of consumption took place during the second half of the eighteenth century, especially in Great Britain and, secondarily, in France, North America and other European countries. However, explanations for how this change came about, and why it occurred in certain locations before others, have fostered considerable debate among scholars in disciplines ranging from history to sociology, anthropology and beyond.[4] As a new object of potential commodification, the ways in which savants and entrepreneurs treated balloons offer insights into the manner in which new thinking about luxury and economics, as well as the role of scientific innovation in the marketplace, had evolved at the end of the age of Enlightenment.

Certainly, the invention of balloons led to a vast array of economic opportunities and aeronauts acted quickly to market balloons, in a variety of forms and formats, to the public. How did entrepreneurs make balloons, this new necessity, available and attractive to consumers? How could the public invest in launches and why should they care to do so? The first aerostatic machines were large and costly affairs that the average person could not possibly afford. Thus, marketers had to reconceptualize the balloon into something that their customers both wanted to buy and which would fit into the price range of a large enough segment of the population so that a profit could be made. One of the ways balloonists accomplished this was by selling launches to large groups of people. In addition, as seen in the next chapter, other entrepreneurs created balloon-related merchandise, such as books, prints and material goods with balloon themes, which they promoted to the general population.

Enterprising individuals with a keen interest in aeronautics began selling balloons and quickly set prices low enough for average, middling-sort consumers to participate. The broad success of ballooning in England and France, just as those regions had greater success with their economies in general, arose from the appropriation of a variety of marketing tools to identify potential consumers

Figure 5.1

and fund launches. While balloonists actively marketed flights in other regions too – Jean-Pierre Blanchard set up subscriptions wherever he went throughout Europe and North America – the breadth of ballooning in England and France owes much to the presence of both a larger middling class, already the target of entrepreneurs, towards which aeronauts focused their advertising, and the fact that this group was interested in purchasing something as fleeting as access to a launch. The number of people able to obtain sufficient funds for their experiments, even if only for a short period, demonstrates the public interest and ability to invest in these scientific events. This chapter examines the methods used by balloonists to fund their launches and especially their use of newspaper advertisements to announce advance subscriptions and ticket sales. In addition, this chapter explores some of the reasons why people chose to spend money on launches; in other words, what relationship existed between audiences, balloonists and launches? The overall goal is to illuminate the possibilities of economic involvement in aeronautics, and science more generally, by a wide group of people as well as to suggest the meanings they might have attached to their investment in this practice or, at least, the meanings marketers thought people might connect to the purchase of balloons. In other words, this chapter explores the means by which savants might market themselves, and science, to the general public and the ways in which people could appropriate ballooning in accord with their own interests.

Marketing Balloons

This is 'the advertising age', announced the headline of a 1785 article appearing in the *Publick Advertiser*. They went on to remark 'on the great number of advertisements in the news-papers including those aimed at people who wanted to sell an estate, get a place at court, show a pig, teach turkeys to dance, procure charity, marry a wife, keep a mistress, buy dogs, nurse children and, of course, to go up in a balloon.'[5] Since balloons large enough to carry passengers were too expensive to sell to individual consumers, balloonists needed to take new approaches to determine those parts of the launch process they could market. Rather than selling the actual balloon, entrepreneurs concentrated on selling the launch, or at least the opportunity to witness and participate in the launch. A large number of tickets were issued, some of which have survived; Figure 5.1, for example, is a ticket from one of Lunardi's ascensions in Lisbon. Of course, once a balloon had taken off and reached a certain point in the sky, it became impossible to control who watched. Thus, balloonists sold other aspects of the launch process leading up to that moment. Proximity to the balloon prior to the launch became important; essentially, people paid higher ticket prices in order to have a seat or to sit closer and in greater comfort. Balloonists also introduced various forms of entertainment alongside the launch that became part of the sales pitch. Music, speakers and the presence of famous or important people all emerged as potential elements of advertisements. Subscription packages, which came with launch tickets, might also include access to pre-launch activities such as the construction and inflation of the balloon, a lecture on aeronautics, and the right to sit in the gondola prior to launch. While many theories for consumer behaviour have been proposed, the focal points here will be, on the one hand, the emphasis aeronauts placed on the entertaining, spectacular nature of ballooning and, on the other hand, the connection between balloons and the Enlightenment.[6] The birth of aeronautics provides an example of how commercially-minded aeronauts marketed a new product in order to attract an audience.

While a dearth of sources prevents direct analysis of why people invested in balloons, aeronautic advertisements provide a wealth of information on what balloonists thought would attract consumers to their launches. Certainly there was a novelty factor. As the latest scientific wonder, people flocked to see and experience this new invention.[7] But it remains difficult to comprehend how people understood this experience and why, specifically, they decided to invest limited resources into accessing launches. The main difficulty in unravelling the mentalities of early modern shoppers lies in the nearly impossible task of accessing consumer thinking in the early modern period.[8] Colin Campbell notes:

> Consumer goods serve to satisfy needs or indulge wants and desires. In addition they may serve to compensate the individual for feelings of inferiority, insecurity or loss, or to symbolise achievement, success or power. They also commonly serve to communicate social distinctions or reinforce relationships of superiority and inferiority between individuals or groups. They can also, on some occasions, express attitudes or states of mind, or communicate specific messages from one person to another. Finally they may be instrumental in creating or confirming an individual's sense of self or personal identity.[9]

The reasons why people may have wished to participate in launches form a wide spectrum of possibilities. Balloons could be assigned a variety of cultural values – such as 'useful', 'enlightened' or 'scientific' – and, as such, could be used by consumers to fulfill a variety of functions.[10] Consumers may have felt that they acquired some cultural capital through their participation in the aeronautic marketplace.[11]

Balloonists presented their wares to the consumer public through a series of printed sources that provided potential purchasers with a rationale for their acquisition. This justification usually played on two sides of ballooning – its spectacular nature and their usefulness whether philosophical, economic and scientific. Of course, the same sources may be interpreted in a variety of ways and we cannot be sure exactly what meanings individuals took from what they read. Advertisement for balloons frequently pushed the idea of launches as spectacles.

With the enormous size of the audience for these events, this is arguably an early form of mass consumption centred on entertainment.[12] Zambeccari expressed happiness in the thought that his launch contributed to the entertainment of the people.[13] This trope received further reinforcement through the number of festivals using balloons, such as the coronations of Napoleon and Louis XVIII in France and George IV in Great Britain, as well as for holiday celebrations such as Christmas. During the French Revolution balloons helped commemorate the fall of the Bastille. Balloons were associated with a party atmosphere and might be launched in the midst of music, food and dancing.[14] Aerial voyages were a 'new spectacle' that appealed to the interests of the 'curious'.[15] Balloonists sold their invention as a great curiosity that people would, of course, want to see close up. In this sense, balloons can be viewed as a luxury, albeit one set at a price more affordable than most luxuries, which were designed to promote pleasure and happiness. All fashionable people would want to take part in this entertainment.[16] Thus, balloonists identified their wares as potential entertainment; this did not require much imagination, but it did mean that aeronauts needed to create a festival-like atmosphere around their launches and pre-launch activities. The increasingly elaborate activities associated with launches are a direct result of this realization on the part of aeronauts.

In addition, balloonists conspicuously identified their launches as experiments which could contribute to humanity's overall understanding of its place in the world.[17] Aeronauts often made a point of detailing what kinds of experimental equipment they planned to take on their launch; meteorological devices appeared most often although sometimes the advertisements remained somewhat vague, referring only to instruments and machinery.[18] The emphasis on the launch as an experiment worked as a method to include the audience in the event; balloonists transformed customers into witnesses for the scientific performance who could then testify as to its veracity and success. Thus, just as the Montgolfier brothers had done at their first launch in Annonay, balloon experiments in general, as well as more specific aspects of the launch like new methods of filling the balloon or of steering it, became opportunities for the public to participate in this new science.[19] This helps explain the vast number of eyewitness accounts published as pamphlets or in the periodical press. As balloons came to symbolize the age of reason, especially with their focus on the conquest of nature and the potential utility of the invention, aeronauts implicitly carved out a new manner in which individuals, for the price of a ticket, could become active members of the Enlightenment.[20] Buying a ticket for a launch, then, might also mean that the ticket-holder could claim to have witnessed an important new discovery relating to aeronautics. Thus, while it remains difficult to identify exactly how participants might have used balloons to fashion their own identity, balloonists could use their customers to create an experience in which the audience verified the success of their scientific work. In this fashion, the presence of the audience directly shaped scientific practice.[21]

To construct, fill and launch a balloon of any considerable size could cost an enormous sum of money. This proved particularly true if the balloon was large enough to carry passengers. While the Montgolfier brothers, Etienne and Joseph, paid for their early experiments out of their own pocket, they certainly did not intend to maintain that method for funding further experiments. Instead, they quickly turned to the state for funding and support. In other words, the Montgolfiers appealed to state patronage for help. The Montgolfiers carefully planned the unveiling of their new invention to coincide with the diocesan assembly of Vivarais in Annonay; this group of men had the appropriate status to act as acceptable witnesses to the launch and sign an affidavit forwarded to the crown and the Académie Royal des Sciences.[22] The Montgolfiers, in part, attempted to garner official accolades and recognition; however, their actions also reflected the economic reality behind the expenses that balloonists might incur.

Before the Montgolfiers fully initiated the process for patronage, however, Barthélemy Faujas de Saint-Fond, an enthusiastic supporter of balloons, called for an alternative form of funding ascensions – namely, the public subscription. For the first launch in Paris, Faujas de Saint-Fond established what he claimed was the 'first national subscription', based in a variety of locations but especially the famous Café de Caveau at the Palais-Royal in Paris.[23] This tactic made sense, at least to Faujas de Saint-Fond, for a variety of reasons. First, no obvious or automatic institutional base for ballooning existed and so no direct avenue for funding requests was truly available. Some academicians expressed interest in ballooning, mostly people like Lavoisier who performed some experiments on hydrogen in support of his chemical reforms; in addition, the Académie Royale des Sciences established two commissions to study the topic.[24] However, none of the early balloonists enjoyed membership in a major royal academy. Neither were they independently wealthy. Although the Montgolfiers, as part of a large paper-making family, were fairly well-to-do, Jean-François Pilâtre de Rozier and Jacques-Alexandre-Cèsar Charles worked as popularizers of experimental physics while in England, James Sadler was the son of a pastry chef and Vincenzo Lunardi a minor government official. Early balloonists came from a variety of walks of life and were not especially well off as a group. From the beginning, therefore, launches were in the hands of people outside the scientific and economic elite who needed to make money from their efforts in order to earn a living.

Initially, of course, subscriptions did target an elite audience. Balloonists used this method to fund even moderately priced launches, such as the one organized by the Abbés Calmet and Touche in Rodez in August of 1784 which cost a mere 1,800 livres to stage.[25] The subscription set up to finance the Montgolfier brothers' 1784 launch in Lyon aimed to raise a total of 4,320 livres; the goal was to enrol 360 people at twelve livres each. Ultimately, they sold 283 subscriptions to 174 people and collected 3,396 livres toward the cost of the equipment. Where indicated, the status of subscribers tended to be fairly elite with counts, barons and other nobles represented along with the local Intendant, lieutenant of police, tax farmers and other office holders. Savants also participated, with ten academicians listed. Women, mostly those married to other subscribers, bought eleven of these. Only around 10 per cent of the subscribers appeared to be of bourgeois or merchant status.[26]

Several balloonists opened equally expensive subscriptions aimed at the French elite. When Pilâtre de Rozier announced his first flight in a Montgolfier-designed balloon he received 1,200 livres from his noble patrons and opened a subscription to fund the rest. The tickets cost twenty-four livres and came with four passes for viewing the launch.[27] Other balloonists adopted this model, such as in a 1784 launch in Dijon. For twenty-four livres the subscriber received four tickets to the launch as well as admission to a second launch to take place some time later. The balloonists invited individuals who purchased three subscriptions to attend all the preliminary work in the construction and inflation of the balloon and gave them the honour of having their names printed, along with a description of the flight, after the experiments were over.[28] Alternately, to help advertise their event, Charles and the Robert brothers put their balloon on display several days in advance, along with the rest of their equipment, a tactic more frequently employed in Britain. This enormously successful subscription may have earned them as much as 50,000 livres. Charles and the Robert brothers clearly collected more than just respect for their efforts. [29]

Quickly, French balloonists began to set their subscriptions at a lower rate, usually three livres or less, in order to allow a larger segment of the population to purchase access to launches. In England, where Zambeccari, in the opening example, had charged five shillings and more, the rate for tickets fell to around one shilling although some tickets remained as high as 2 *s* 6 *d*. In other words, most aeronauts tried to garner patronage from a larger number of people each paying a small amount. This was an essential shift in ballooning, and popular science in general. By targeting a broader audience, these balloonists placed their status, economic livelihood and future in the hands of the public in ways that their elite counterparts did not wish to do. Not all balloonists, however, did this. Zambeccari, for example, tried to limit the number of people at his launches. He would not announce the day or time of the launch but instead printed that information on the tickets; in addition, he pleaded with ticket holders to 'forbear communicating' that information to the public in order to 'prevent the inconvenience of too numerous an assemblage of spectators'. Rather than an effort to focus on science, this seemed equally an attempt to prevent a riot in case of failure and to ensure that all those who saw the launch had paid for the privilege.[30] However, there were only a few balloonists who could afford to eschew public support. Thus, almost immediately after the advent of ballooning, most aeronauts began selling this new invention to a ready and willing public rather than focusing solely on the top end of the social and economic hierarchy.

Advertising Balloons

While balloonists may have had to invent the field of aeronautics, they did not have to reinvent the wheel when it came to the marketing of scientific phenomena and instruments. Entrepreneurs had sold science and science-related objects to consumers for some time. Individuals enrolled and participated in public lecture courses as well as purchasing science books, broadsides and other commercial ventures related to just about every branch of science that appeared consistently throughout the late seventeenth and eighteenth centuries.[31] Most of this exchange targeted individuals rather than groups. Public lectures proved the exception, of course, since popularizers leading these courses encouraged a larger audience given that the more people attending translated into greater profits for the instructor. Similar to launches, public lecture courses might require a considerable outlay of capital in terms of equipment and supplies. Even after a lecturer established himself (or, more rarely, herself), maintaining and updating equipment and supplies continued to necessitate a steady income. But the scale here paled in comparison to the expenses associated with aeronautics. Not only did balloons cost a lot of money to construct and inflate, but they also tended to break down at the end of almost every launch since landing balloons proved more punishing on them than the take off. Even if an aeronaut could use the same balloon multiple times with, perhaps, only minor repairs, it still required inflation for each use. Thus, balloonists needed to turn to alternative methods of funding their work.

Just as balloonists had models to follow in presenting their work to a paying audience, so too did the audience have experience in appropriating science. In general, people did not assume a passive stance in their relationship with knowledge acquisition; instead, they selected those aspects of scientific culture that appealed to them and felt free to ignore the parts they found less interesting or useful. As Jan Golinski has shown in the case of barometers, people who bought scientific equipment often did so with specific notions of how to adapt or alter the instruments to their own ideas and interests. Such consumers of science, argues Golinski, might also interpret instruments in ways antithetical to the ideas attached to them by their creators.[32] Similarly, purveyors of balloons probably had concrete ideas about how viewers should understand and interpret their work. On the one hand they saw the launch as entertainment and spectacle; on the other hand, some balloonists certainly wanted people to believe they conducted serious, enlightened scientific work, and possibly some of them did. These individuals may have hoped to create an image of themselves as savants as well as, or perhaps rather than, showmen.

Consumer behaviour during the age of the Enlightenment encouraged public participation in various cultural activities. A number of individuals and groups had already employed subscriptions as a money-raising tool prior to the advent of ballooning. Authors and publishers often sold books in this manner as a way to ensure that most of the copies from any given print run would already have a market. The publishers of the first edition of the *Encyclopédie*, for example, operated by way of subscription. This practice expanded to include the creation of subscription libraries and reading societies. Certain kinds of charitable institutions, such as hospitals, were founded through subscriptions.[33] In England some subscriptions were taken up to raise funds for the building of ships. Some cultural events also might receive public sponsorship through this means. Mozart, for example, dared to turn his back on noble patronage for a series of three publicly-funded concerts in 1784.[34] Thus, balloonists, on the one hand, appropriated a technique common in the Enlightenment to fund cultural activities. On the other hand, they were not operating against a normative standard of noble or royal court patronage for their particular event. On the contrary, as purveyors of a new commodity, once aeronauts introduced subscriptions, something that happened almost immediately in the history of ballooning, they became the norm. As such, balloonists found their launches almost solely in the hands of the paying public rather than at the mercy of state or noble sponsors. Generally, ticket sales were modelled on other similar entertainments such as a plays and music concerts; this added to the public control of the event. This made the funding of such projects a rather unique activity in the late eighteenth century. Some individuals did critique subscriptions; John Jeffries, a doctor and companion to Blanchard across the English Channel, thought they were a 'tax on the curiosity of the public' and lamented that no better means had been established for funding launches.[35] For the most part, however, balloonists successfully and consistently used subscriptions and tickets to pay for the efforts.

Eighteenth-century consumers, however, could not turn automatically to a specific set of merchants in order to satisfy their desire for aeronautica. On the contrary, balloons did not fall immediately to one specific subset of the market economy. Instead, balloonists and the people who manufactured and sold balloons and balloon-related merchandise emerged from a variety of areas and included a civil servant, several scientific lecturers, an umbrella manufacturer, a maker of artificial flowers and a confectioner. Thus, the public did not have an obvious place to go if they wanted access to balloons; no 'Balloon Street' existed as a counterpoint to 'Grub Street' and the profession of 'balloonist' was up for grabs to whomever wished to claim the title. Without a tradition to follow, aeronauts had to reach out to their potential audience and help them navigate this new terrain in ways not necessarily required of other, pre-existing merchants. To help their audience identify themselves, balloonists turned almost immediately to newspaper advertisements which were also undergoing extensive changes in form. Alongside a rapid growth of consumption came changing attitudes toward competition and marketing.

Prior to this transitional period, advertisements did not necessarily focus on enticing customers or trying to lure potential consumers away from other proprietors. In both England and France, at least initially, advertisers felt 'little desire to puff the wares' and there was 'none of the subtlety found in modern marketing which more often than not tries to create a feeling of sympathy between the potential customer and the product without listing the latter's qualities too overtly, while aiming it precisely at a clearly identified social group'.[36] However, the new commercial advertising, much of which targeted the emerging middling classes, did begin to engage in exactly this kind of behaviour. In other words, 'advertising' had generally implied the public nature of a thing in the earlier period while later definitions focused on it as a 'way of exerting influence on the public for commercial ends', which is to say it became appropriate for advertisers to use newspapers to try to attract customers.[37] As marketers of a new product, aeronauts advertised in such as way as to attract a market, something not accomplished automatically even with an invention as innovative and different as balloons. Of course, not all references to balloons in the periodical literature of the day were commercial. On the contrary, a considerable number of articles appeared that simply announced the impending launch of a balloon or reported on the relative success or failure of the experiment. Frequently, local newspapers included information on launches from other countries, especially if they were performed by particularly famous aeronauts or, predictably, if some disaster or problem attended the launch. However, commercial advertising of balloons played a significant role in the eighteenth-century press; the press flooded the market with almost daily news reports on this invention, those involved in its development and details of all launches.

As John Styles has argued, new products require some sort of product definition in order to become successful.[38] So, how were balloons marketed? A great deal of information can be gleaned from advertisements in newspapers about the nature of this product, its intended consumer base, and the reasons why consumers should be interested in it in the first place. When it came to the material forms of consumer culture, entrepreneurs simply added balloons to already existing products. Thus, the profusion of balloon hats, games, liqueurs, furniture and so on. Ascensions, however, were unfamiliar and entrepreneurs needed to establish a market for them. The 'newness' of the product had to be 'reconciled with consumers' pre-existing experience, knowledge, and expectations'.[39] Marketers created an initial need for balloons and strove to maintain interest in this new product over time. While a few of their techniques appeared ubiquitous, some marketing strategies differed.

Initially, as one might suspect, aeronauts felt only a minimal need to advertise balloons. Their immense popularity and novelty ensured that this new invention dominated newspapers, letters and conversations throughout Europe and the Americas. As Faujas de Saint-Fond wrote, amateurs in physics found themselves intensely occupied by balloons and, when he announced the subscription for the first experiment in Paris, little effort seemed necessary to gather enough participants to cover the costs.[40] For this launch, of the hydrogen balloon created by Charles with the help of the Robert brothers, Faujas de Saint-Fond publicized a subscription with the hope of raising 10,000 livres.[41] As Antoine de Rivarol noted, tickets simply indicated the 'place, the day, and the hour' of the launch. The owner of the Café de Caveau acted as treasurer for the subscription; he received so many requests for tickets, reported Rivarol, that he claimed to have never been more financially happy.[42] The *Courrier d'Avignon* also reported on the success of this subscription. 'Princes, ministers, men of letters, [and] citizens of every condition [tous états]', the author of the article announced, were impressed enough by the enterprising work accomplished by these physicists to purchase tickets.[43] The tone of the advertisements, however, was not particularly commercial. Instead, it was clear that the goal was to fund a scientific experiment of national importance rather than pay for a good, a service or an entertainment. In any case, the call for assistance to fund this first Parisian launch resulted in an abundance of support leading it to become oversubscribed, although that was certainly not a problem. Faujas had aimed for 10,000 livres, but rumours suggested that the amount raised came closer to 50,000. This resulted, according to one report, in some financial wrangling among Charles, the Robert brothers, and the owner of the Café de Caveau, perhaps on how to divide the surplus money.[44]

Over the first year or so of launches, other calls for subscriptions were equally straightforward. Typically, several locations were identified for the purchase of tickets. Faujas de Saint-Fond had used the centrally-located and popular Café Caveau at the Palais-Royal in Paris as the main location for his subscription. The balloonist Campmas sold tickets throughout Paris for his April 1784 launch, mostly from cafés including the Café de Foi, another Palais-Royal institution and rival to the Café Caveau. But Campmas sought subscriptions from a broader geographical area than had Faujas de Saint-Fond; tickets for Campmas's launch could also be purchased north of the city, near the Temple, as well as at two locations on the Left Bank, near the Sorbonne and near the Odéon.[45] The Robert brothers established a subscription for their launch from the Tuileries garden in November 1783. The emphasis in their advertisements dealt almost solely on logistical questions concerning how ticket holders should enter the garden.[46] A chemistry professor in Nancy named Nicolas opened a subscription, set at the increasingly standard three livres per ticket, for his launch in late 1783. Tickets were available from a M. Willemet, an apothecary; he claimed the subscription was only to 'cover the costs of the experiment'. It does not seem as if Nicolas had definite pecuniary ambitions; he simply tried to defray the considerable expenses associated with ballooning.[47] Organizers of the early launches in France, therefore, do not appear particularly mercenary in character, nor did it require much salesmanship on their part to lure customers. Instead, the balloon acted as the chief attraction in these notices. Aeronauts apparently did not consider any other enticement necessary to get people to subscribe.

In London, advertisers initially approached their task in a slightly different manner than their French counterparts. For one thing, the first balloonists offering launches in Britain did not speak of subscriptions even though they had that option. Instead, when Zambeccari and Biaggini launched their balloons in November and December of 1783, the advertisements talked only of entrance tickets. The implication here is that the balloons would be launched no matter how many people paid to attend; but they also expressed no worries that enough people would buy tickets to fund the project successfully. In addition, the British ads created a larger market for ballooning by also selling access to the balloon in advance, something the French tended to include in the overall price. Aeronauts in Britain sold tickets for these viewings separately from those of the launch itself thus enabling merchants to reap greater profits from the event. Consumers could pay to watch the construction of the balloon, as well as simply to see it in its inflated state, but in the comfort of an indoor environment. Then they could buy tickets to view the launch itself. Thus, Biaggini constructed and launched a balloon in late November 1783. Interested persons could purchase tickets for this early British launch from Biaggini at a well-known exhibition hall called the Lyceum, at his house, and from a pub called the Swan. The launch itself, overseen by Zambeccari, took place on 25 November 1783 on the grounds of the Artillery Company. The extent to which Biaggini advertised the imminent launch of this balloon caused some comment; the anonymous author of a description of the launch in the *Gentleman's Magazine* noted that the event took place after 'repeated notice given by advertisement in the publick [*sic*] papers'.[48]

The success of this venture encouraged Biaggini to construct an even larger balloon. He published an advertisement to that effect in the *Morning Herald and Daily Advertiser* as well as in the *Gazetteer and New Daily Advertiser*, and the *Morning Post*. Biaggini addressed these identical ads to 'the Nobility and Gentry' and let it be known that his 'Grand Aerostatic Globe', totalling sixteen feet in diameter, could be viewed at the Lyceum every day, from 10 am until 4 pm. Visitors could expect a well-aired room and to find fires lit against the chill winter weather. The exhibition, which cost 2s *6d* per person, remained open from 6 December 1783 until 24 December 1783; after that, Biaggini said, he would announce a launch sometime in the near future.[49] Not all British balloonists followed this option of dividing the event into its component parts and charging people at every stage. James Dinwiddie, who launched the first balloon in Dublin, Ireland in December 1783, focused just on the launch and did not offer any additional opportunities for viewing the balloon or, from another perspective, of making additional money out of the experiment. He simply announced a location, wrote that seats were available, and set a price of one shilling.[50]

The first few months of ballooning, then, required little effort on the part of entrepreneurs to sell the idea of the launch. Patrons and customers appeared quite willing to step forward to view this novelty, whether at the launch or in the various stages of its construction, and required little encouragement to spend their money. Significantly, these early calls for subscribers appear more similar to appeals for patronage than attempts to attract customers. Although other aspects of science had undergone commodification by this time, launches did not automatically become the purview of consumers; instead, the possibility still existed that the wealthy, the scientific elite or the state would control the direction and practice of ballooning. However, the chance for this quickly passed as subsequent calls for subscriptions began to target an ever larger potential customer base. At the same time, the general trend in advertisements earlier in the century, to state the nature of the object for sale and to not dwell on or attempt to convince potential consumers of the need to purchase certain wares, underwent adjustments. Advertisers turned away from such passive methods and turned to more proactive advertisements designed to garner attention for their products. Thus, at the same time that advertising underwent a transformation in purpose and methods, so too did balloonists begin to target a larger group of people. Balloonists could not simply rest on their laurels; instead, the sale of launches required, in the years that followed, more effort and justification on their part. As the spectacle of balloons became common, the need to sell the specific aspects of a particular event grew more necessary. Thus, within a year or two of their invention, many aeronauts began emphasizing the unique qualities of their launches over those of their competitors as some of them began to envision ballooning as a commercial venture.

Throughout 1784, advertisements for launches typically exhibited a tendency to become more elaborate in their offerings. This did not occur uniformly and some subscriptions remained very straightforward. A savant named Levêque in Angers, for example, launched a series of unmanned balloons from that city. His subscriptions cost three livres each and apparently came without any additional inducement, although the subscribers' names did later appear in print as patrons of a particular launch.[51] However, in general, it was no longer sufficient simply to announce that a balloon would take off. Now, aeronauts added embellishments to attract potential customers. Two members of the Académie de Dijon, Gattey and Mégnié, along with a third person named Henry, proposed in March 1784 the launch of a balloon which, they felt, would ultimately be useful to the government, commerce and the sciences. Subscriptions cost twenty-four livres but came with four tickets to the launch that allowed individuals to assist in the launch process in a manner not made explicit. However, in addition the subscribers would have access to a second experiment at some yet-to-be determined date, thus bringing the price per person for each launch down to three livres. People who purchased three subscriptions were invited to the preliminary experiments and their names were placed at the head of the journal which would print the results of the experiment after the balloon landed.[52] Unfortunately, they failed to entice enough people to subscribe; they needed 10,000 livres to launch their balloon and, therefore, required 417 subscribers; but they apparently never reached their goal.[53] These aeronauts had set their sights rather high with a relatively expensive subscription; even though the tickets came with a number of amenities, the cost certainly limited their consumer base. A similarly high priced subscription appeared in Avignon in April 1784. There, individuals could purchase a subscription for 120 livres which enabled them to examine the balloon twice prior to launch and to visit the workshop in which the balloonists had constructed the machine. Clearly, the target audience was very well off.[54]

Other French balloonists tried alternative tactics and set their prices considerably lower. In Toulouse, a self-proclaimed 'amateur' balloonist emphasized the high price of launches along with his own inability to cover all the costs. He sold his tickets for a mere three livres each and hoped to raise 6,000 livres; lest the public accuse him of greed, he assured the readers that he would faithfully account for all the money.[55] Campmas made sure to point out his innovative work in steering balloons when he announced his launch in the spring of 1784. Thus, although his subscription cost the usual three livres, there was always the hope that subscribers would be funding someone who had overcome the single greatest problem facing the science of aeronautics. In this way, he linked his launch more explicitly to the scientific side by concentrating on proving its utility.[56] By the fall of 1784 Campmas began to add further incentives to his advertisements. In addition to his claims regarding his ability to steer the balloon, he also began offering details about the globe itself. In October, he claimed he would launch a balloon shaped like an 'aerostatic tower' sixty feet high that could carry six people. Potential subscribers could purchase tickets at six different locations throughout Paris, mostly in cafés.[57]

In England, some balloonists also started emphasizing new details about the balloons. Of course, as with France, not every balloonist in Great Britain operated under the system of charging admission for both viewing the balloon in advance and watching the launch; nor did every balloonist take advantage of lures to draw in customers. After his successful efforts in Dublin, Dinwiddie exhibited and launched a balloon in January 1784 from Bath without calling undue attention to any special aspects of the globe other than to describe it as 'elegant'.[58] James Tytler of Edinburgh opened a subscription for his launch during the summer of 1784 and apparently had trouble collecting the money. He announced that a few of the subscribers had yet to pay and that the entire enterprise was still under funded. To convince people of the necessity of buying a ticket, he did announce that during some initial test runs the balloon would be kept on a string both to prevent its escape and so that 'its ascent cannot be seen by any person except those admitted' into the viewing area, a local garden.[59]

However, British balloonists increasingly looked to maximize the amount of money their earned from their balloons and developed advertising strategies designed to lure larger numbers of customers. Biaggini, for example, began to differentiate aspects of the launch process. Initially, the division was between simply between the launch and anything that happened prior. Biaggini, however, created a third moment for which people could purchase tickets. Potential customers could visit the Pantheon to see the balloon, for the price of 2*s* 6*d*; or they could pay twice as much which also allowed them to attend the inflation of the globe.[60] This practice continued well into the nineteenth century when an anonymous balloonist in New York City displayed his 'grand inflammable air balloon, (in an inflated state) superb car, richly variegated canopy, the flags and all the necessary appendages appertaining to an ascension' all for fifty cents. Apparently he had tried to launch the balloon earlier, but failed and was using this to raise money until he could set another date for an ascension.[61] Other aeronauts concentrated on introducing new themes to the construction of their balloons. Some advertisements, for example, lauded the connection between balloons and the exotic, a common theme in Enlightenment Europe.[62] Thus, in April 1784 a balloon was put on display prior to launch that was thirty feet high, shaped like a pyramid, and covered with 'caricatures, hieroglyphics, and inscriptions'. The next month, exhibitors presented a balloon at the Pantheon based, so the advertiser claimed, on an entirely new plan and was the 'largest and most curious of its kind that hath hitherto been executed'. It was designed to represent a 'Chinese Temple, superbly decorated with columns, a gallery, &c., and appears, in the operation of filling it, as rising instantaneously out of its ruins, and floating, in a majestic manner, in the body of a thick cloud'.[63] Zambeccari issued a lengthy prospectus for a 'large aerostatic globe' that distinguished between potential subscribers. Those who purchased a one-guinea ticket could bring three people to the launch with them and could arrive either on foot or in a carriage. Those who bought the half-guinea tickets could only bring one person and had to arrive on foot. In addition, he set up 'separate Accommodations in the Place for Persons in Carriages, and for Persons on Foot; the latter will have Seats round the Globe, which will be for Subscribers only, in order to prevent their being incommoded by the Populace'. Subscribers could also view the balloon in advance with the one-guinea ticket holders allowed in four times and the half-guinea patrons admitted twice.[64] While Zambeccari clearly intended to maximize his profits by keeping the riff-raff out, he also offers an indication of how ballooning became socially stratified. The more you paid, the closer you got to the action and the more comfortable your surroundings.

Some balloonists assumed an even more exalted demeanour regarding their efforts, a posture that could backfire. The Chevalier de Moret placed his launches within a more intellectual (or perhaps spiritual) framework in addition to some visual lures. People who paid the one shilling admission fee to view his balloon would see a balloon 'suspended by Omnipotence, floating in the incomprehensible infinity of eternal space!!!' It was designed to appear as solid gold and was 'richly decorated with artificial flowers'. As an additional enticement, individuals who bought a ticket would also be given, gratis, a 'beautiful print of a balloon, and of Monsieur de Montgolfier filling it.'[65] In August 1784, de Moret announced his intention to conduct a manned ascension. In addition to the usual tickets, de Moret had also advertised, in advance, engravings of the balloon. Individuals who bought copies of this print should have counted themselves the lucky ones since de Moret failed to fill the balloon, much less launch it, resulting in a riot and the destruction of the globe.[66] Having emphasized the lofty nature of his experiments, when it failed, the people and the press viciously took him to task.

When Lunardi, in August 1784, announced the imminent launch of his balloon from the Chelsea Hospital Garden he noted part of the profits would go to those impoverished individuals who lived at the hospital.[67] In advertising this launch, purporting to be the first manned flight in Britain (he, and many others, ignored the attempts by Tytler in Edinburgh), Lunardi provided some extra details of the balloon, pertaining to its description and the attached wings with which he planned to steer it; but these first advertisements were, in general, straightforward. Admission was set at one shilling although those willing to pay more could gain additional access. Tickets at five shillings, half a guinea and one guinea entitled the bearer to 'admission into the Gardens of Chelsea Hospital on the day the Balloon is to be launched, and also to a sight of it before it leaves the Lyceum'.[68] Joseph Banks, the president of the Royal Society of London, was one of the first to subscribe, but overall the launch was an economic disaster.[69] After the riot surrounding Moret's failure, the head of the hospital decided not to risk it and withdrew his permission to hold the launch at the hospital. Lunardi scrambled for a different location and finally received permission from the Royal Artillery Company to launch from their training ground.

Whether from disappointment at Moret's failure or confusion over the changing days and locations, Lunardi's launch was disappointingly attended in spite of his advertising blitz. A reporter for the *Morning Post* wrote that 'about five hundred people paid for admission into the Artillery Ground yesterday; very few of whom sat in the half-guinea seats and still fewer in the guinea ones; on a moderate computation, therefore, it is supposed the money taken could not amount to much more than' £150.[70] People living around the Artillery Grounds also hurt Lunardi's profits from this launch; the *Morning Post* chastised these individuals noting that they

> were sorry indeed to observe that such general advantages had been taken by the neighborhood, of farming their windows, and for benefits which were due only to the novelty and spirit of the enterprize; we were the more sorry, as the ground which should have been liberally offered was, under a false idea of generosity, *upon the present occasion*, most unhandsomely rented.[71]

This critique also appeared in the *Bath Chronicle* which added, in agreement with other periodicals, that of the estimated 150,000 who attended the launch, only about 500 paid money to Lunardi 'who at enormous expense, and certainly hazard, made the philosophical effort. Most of the people in the neighbourhood', the newspaper added, 'profited from the curiosity of the public. Their houses were scaffolded from top to bottom; and all those windows which Mr. Pitt's commutation tax had blocked up, were re-opened for the purpose of accommodating spectators at half a guinea and five shillings a head'.[72]

Lunardi's inability to reap his proper rewards even resulted in the formation of a public subscription after the launch, with the avowed intention of rewarding his 'merit' and 'as a recompence for the advantages' taken of him by the people living around the Artillery Grounds.[73] The post-launch viewing of the balloon at the Pantheon also proved enormously lucrative. One newspaper suggested that as many as two thousand people attended on 1 October 1784. The Pantheon 'seems to have become the fashionable lounging place for all the beauties in town'. Attractions included the balloon, 'suspended to the Top of the Dome' along with Lunardi 'and his poor fellow Travellers the Dog and Cat who still remained in the Gallery to receive the visits of the curious'.[74] In addition, a 'full Band of instruments [was] provided for Minuets, Cotillions, and Country Dances' while some rooms catered to those who wanted refreshments. Lunardi set the price for attending the dancing, which lasted from eight in the evening until midnight, at five shillings, considerably more than the one shilling needed just to view the balloon. Nonetheless, as Betsy Sheridan reported, everybody 'gives their shilling to see it'.[75] We can see Lunardi targeting specific segments of the population with his pricing and seating structure. But the difficulty in gaining customers also seems apparent. In New York City a balloonist named Charnock claimed to have 'sustained a loss the night of his benefit' and so pleaded with the public to buy more tickets, at fifty cents each, to attend a second launch.[76]

Selling a launch had become an elaborate process by 1785. There was a danger that people would not find ascensions enough of a draw on their own. In fact, Louis Philippe de Ségur wrote in his memoirs that balloons had 'become so common, that [a launch] scarcely excites any farther curiosity'.[77] A broadside from 1785 even showed a picture of an itinerant aeronaut 'out at the Elbows' roaming from town to town begging for money to inflate his balloon.[78] Launches required additional inducement to lure customers while at the same time balloonists developed increasingly new and novel balloons. In some cases, the shape of the balloon made the difference. Johann Karl Enslen formed his balloons into a variety of pleasing shapes including, of course, Pegasus as well as a Nymph and a woman with a balloon hairstyle. He exhibited these figures across Europe. In France he joined forces with the Ruggieri brothers, a famous family of pyrotechnicians, to launch his balloons prior to a fireworks display. Seats cost three livres for a seat in the gallery or a loge, or one livre ten sous just for entrance into the garden.[79] In London Enslen performed his spectacle at the Bermondsey Garden in Southwark where the best seats cost 2*s* 6*d* and the less expensive places just one shilling.[80] Enslen reappeared in Paris and London occasionally over the next several years as he toured the continent with his balloon figures.[81] These unmanned launches had considerable success and were potentially more profitable although less dramatic than manned flights. The profit margin would have been higher since the balloons were smaller and reusable. But without the thrill and risk of a passenger, Enslen's launches may be considered largely aesthetic events.

As customers became inured to the basic launch, a larger number of perquisites for subscriptions appeared at this time. In the United States, Blanchard, along with his American helper Gardiner Baker, offered an ascension in 1796 for which, 'each possessor of a ticket will have free access to the [balloon], for at least fourteen days before the ascension, at least two hours in the forenoon, and two in the afternoon, and have an opportunity of examining the Balloon, and all the necessary apparatus for lifting and embarking the same'. Baker, who wrote the advertisement, claimed all these events 'will be a most pleasing gratification to the beholders.'[82] In France, Alban and Vallet announced their 1785 launch from the town of Javel in a series of advertisements. They said that they were circulating a prospectus; anyone who purchased the prospectus, at three livres a copy, gained one-time admission to see the balloon and its accompanying apparatus in advance of the launch. They also offered a more extended and very expensive course on ballooning, priced at five gold Louis, which came with many amenities. Subscribers could stop by ten times to see the balloon during the month prior the launch, could climb into the gondola, received a complete explanation of how balloons worked, and had seats to watch the launch.[83] Alban and Vallet recognized the need for more elaborate offerings, but also still hoped to attract an elite audience. Their rapid disappearance from the balloon market suggests they may have failed to garner enough customers.

In Britain, balloonists in 1785 also included additional lures in the advertisements. Richard Crosbie, advertising a flight from the Ranelagh gardens in Dublin, began offering the balloon for viewing in advance. Subscribers to the launch could view it as often as they wished while non-subscribers paid one shilling.[84] In May 1785, Stuart Amos Arnold advertised the unfortunate necessity of limiting access to the construction of his balloon. So many people tried to watch that the workmen complained of all the interruptions. As such, Arnold limited viewers to those who had paid one guinea for tickets to the launch; those who just wanted to watch the balloon being built could, however, pay a one-time fee of one shilling. The one-guinea tickets admitted holders to all the prior exhibitions, a morning excursion on the day of the launch, as well as a seat for the main launch which, if all went well, would take Arnold and his two compatriots, his son and George Appleby, to Paris. For the ten-day period immediately prior to the launch, and after its complete construction, the balloon would be available for viewing free of charge for subscribers and for *2s 6d* for non-subscribers. This exhibition included a band for the entertainment of the spectators.[85] The inclusion of music and the development of a range of privileges for ticket holders indicate a continued interest in forming new methods for attracting customers. Blanchard also included music for his American launch; it was 'played during the time of inflating'.[86] For one of his London launches Blanchard created a fairly elaborate system by which the original subscribers, along with holders of half-guinea tickets, had access to the apartments and had seats that 'will be under tents' with 'proper shades'.[87] In 1817 music was a part of Mr Humbert's aerostatic display at the Vauxhall Gardens in New York City. A band was set to play 'a variety of patriotic and enlivening airs' in between launches while the globes themselves were decorated in 'the colors of our country' and during the ascent people would wave flags.[88] Adrian Napey apologized, in 1818, for being unable to engage a band of music 'owing to the troops turning out on Monday'.[89]

By the late 1780s, a few other changes were made to advertisements that balloonists hoped would help get people to invest in their launches. In some cases, a new price structure was introduced and the overall cost reduced. Monsieur Brun opened a subscription in 1787 in which he had a two-tiered price structure, a tactic already in use in Great Britain. Consumers could either pay a higher price, three livres, and get seats close to the launch; or they could pay a mere thirty sous for seats farther away from the action.[90] The latter price was half that of the norm and certainly would have allowed a larger section of the population to buy tickets. As Brun was working in the French provinces, this price structure may also have reflected the ability of people in his area to afford paying the price of admission, but it might also have been an effort to include yet more people, further down the economic ladder, into the balloon marketplace. Peter Carnes, whose launch ultimately failed, also had a two-tiered price structure for his attempt in 1784. He charged $2.10 depending on where you sat. In addition, he hired people with firearms to guard the field; these guards were authorized to take 'the life of any person who attempts to force his way into the field, as no person will be admitted buy by presenting a ticket at the gates'.[91] In 1803 J. West opened a subscription for a launch in Providence, Rhode Island for which he guaranteed satisfaction or your money would be returned. Tickets for gentlemen cost fifty cents but they could, in turn, invite 'Ladies and their Children gratis'.[92] Adrien Napey conducted a launch in 1818 for which 'no price is fixed'. He wanted to 'collect something for the support of a numerous and helpless family' who 'are now left entirely destitute'. Thus, he left the price up to the generosity of those who attended. At other times, however, Napey charged fifty cents for attending one of his launches.[93]

Balloonists, then, slowly increased the number of opportunities for spectators or added elements to the launch that may have made it at least seem like subscribers and ticket holders were getting more for their money. After the initial rush of interest in watching a balloon take off, aeronauts deemed it necessary to add inducements such as the right to view the balloon in advance, have seats for the launch, learn about the physics of ballooning, have music and so on. However, other balloonists also tried to increase the spectacle of the balloon flight itself; in other words, they began to place increased emphasis on the spectacular and the overall theatrical entertainment value of the launch.

A voyage advertised in Ireland during the summer of 1785 in which Mr Durry proposed to launch a balloon, accompanied by his sister and a dog, illustrates another method of tempting new customers. At the right moment, Durry claimed, the dog would be suspended over the side of the gondola, wearing nothing but a parachute, and dropped.[94] This duplicated an earlier experiment done by Blanchard who, on 3 June 1785, also experimented with a parachuting dog. The dog lived to tell the tale and Blanchard repeated this experiment several times with dogs and a cat although he was less successful with a sheep.[95] Dropping animals in parachutes became a common lure in advertisements for the next decade, topped only by the descent of humans in parachutes, something that started in France in 1797 with André-Jacques Garnerin.[96] While Garnerin was the first successful human parachutist, others had considered the idea. After a particularly disastrous flight, Arnold advertised that in his next attempt he would send one of his friends down in a parachute. This flight, however, never took place.[97] In the United States, Mr Guille advertised his intention to jump from his balloon wearing a parachute during a launch in Charleston, South Carolina. Tickets were only twenty-five cents with children admitted for half price.[98] Mr Blunt did the same experiment in Massachusetts in 1800 for the same price.[99] Alternately, Tétu de Brissy offered a different set of incentives that he thought might make his launch more desirable. In 1786 he announced that he would establish an endurance record and stay aloft for twenty-four hours; unfortunately, his balloon did not cooperate and he only lasted five hours. During the French Revolution Tétu de Brissy returned to ballooning with a launch that included himself astride his horse.[100]

The Ruggieri brothers decided to link launches with pyrotechnics. Initially, they just had balloonists, such as Blanchard and Enslen, come to their gardens and launch balloons. By 1789 they were launching their own balloons with fireworks following the launches. In one instance the Ruggieri family appealed to Parisians' sense of public duty by advertising that the proceeds from a series of launches in 1789 would go to charity.[101] A subscription in Providence, Rhode Island in 1790 also included a caveat for charity. If the subscription for the launch raised forty dollars, ten dollars would be given to the poor.[102] By the early nineteenth century, predictably, some balloonists also set off fireworks from their balloons. Marie-Madeleine Sophie Blanchard frequently performed launches combined with fireworks displays from the Tivoli Garden in Paris although this practice ended disastrously.[103]

While a number of balloonists came and went quickly, unable to establish themselves in this new market, some, like Marie-Madeleine-Sophie Blanchard, transformed into professional aeronauts. This was no small task and, for the most part, people who tried this failed. Nonetheless, the attempt to create a new job indicates the potential some people saw in the developing science of ballooning. As with many professions at this time, aeronautical entrepreneurs tended to run their businesses as a family affair. Jean-Pierre Blanchard's second wife (he abandoned the first to pursue his career), Sophie took over the business after his death while Etienne-Gaspard Robertson worked with his son Eugene, as did James Sadler with his two sons, John and Windham; Jacques Garnerin teamed up with his wife, Jeanne, and niece, Elisa. Elisa Garnerin and Marie-Madeleine Blanchard both had fairly lengthy and successful careers.[104]

Blanchard provided the model for an active, if not exactly successful, professional aeronaut. He set out quite early to make a living from balloons and worked consistently from 1784 up until his death in 1808. During that time he made over sixty launches while simultaneously earning a reputation as someone more interested in money than science.[105] He marketed himself quite vigorously; he claimed to have invented ballooning prior to the Montgolfier brothers (based on his failed attempt to launch a flying boat in 1781) as well as the parachute.[106] He often commented on the number of firsts he accumulated; he was the first to fly across the English Channel and the first man to ascend in a balloon in a number of countries including the first flight in a German-speaking area

(3 October 1785 in Frankfort sur Main), in Belgium (20 November 1785 in Gand), in Switzerland (5 May 1788 in Bâle), Poland (30 May 1789 in Warsaw), and the United States (9 January 1793). In general, he used the local papers to announce the impending launch and establish a subscription; when it filled he would arrive and conduct the launch. In this manner he travelled throughout Europe for around twenty-five years where people, mostly, lauded him for his daring feats. Nonetheless, at his death he was not only poor but in debt and his wife decided that she would continue in his footsteps in order to pay off his creditors. She accomplished this and, in the process, achieved her own fame. When Jacques Garnerin's balloon, launched to celebrate the coronation of Napoleon, allegedly managed to sail across the Alps and land near Rome on the tomb of Nero (an apocryphal story that is too good to be true) Blanchard took over as the emperor's official balloonist. She continued to work for Napoleon and, after his fall, became the official 'aeronaute of the Restoration', a position she held until her death.[107]

Blanchard believed that his prices were set so moderately that 'no person will be unable to purchase a ticket, and thereby have an opportunity of examining the balloon, and everything necessary to its preparation and equipment as minutely as they please'.[108] Nonetheless, tickets did not always sell well; Blanchard believed that balloon subscriptions fared worse than any other kind of subscription for the simple reason that for many launches people could so easily view them without paying, as long as they did not mind skipping out on all the pre-launch events and perhaps even miss the exact moment of take-off; in a typical launch, Blanchard thought, 'out of more than four or five hundred thousand spectators, there have not been more than one thousand subscribers'. Thus, he claimed, he had never been compensated fully for his efforts.[109] Others, however, were quick to suggest that Blanchard exhibited undue greed. John Jeffries, his partner on the flight over the English Channel, reported that Blanchard stuck him with the bills for their journey and flight, and then tried to prevent Jeffries from joining him in the balloon by wearing a concealed, and heavy, girdle to make it seem as if the balloon could not carry both passengers.[110] One British critic simply suggested that Blanchard catered to the rich and should have set his prices at a lower level.[111]

While Blanchard made extensive use of advertisements to maintain public interest in his work and announce impending launches, he also wrote treatises after many of his launches detailing what had occurred. Many of these were popular enough to be translated (several appeared in both French and English).[112] Blanchard filled these accounts with sometimes fantastical details; in one he claimed to have dropped a flag from his gondola but then released hydrogen from the balloon and descended so fast that he recovered the flag before it hit the ground.[113] Although accusations of greed against Blanchard were not unique he proved an especially popular target for critics due to his widespread fame, undisguised commercial attitude toward ballooning, and penchant for self-aggrandizement.

*****

In sum, while the commodification of balloons certainly offers some insight into the nature of consumerism at the end of the eighteenth century, they might also fit into the rubric of mass consumption and act as an early, if limited, example of what would come to dominate society at the end of the nineteenth century. The spectacular nature of launches and the number of people who could participate, sometimes numbering in the hundreds of thousands, gestures to a scale of consumption rarely seen prior to the modern age. Even if not all the spectators purchased tickets to launches, or bought balloon paraphernalia, their mere attendance indicates the consumption of a cultural moment in much the same way that some individuals simply went to department stores in order to experience the environment and enjoy looking at the available goods.[114] The willingness of aeronauts to target as broad a potential customer base as they did indicates a general trend in that direction.

The price of tickets and subscriptions remained reasonably stable throughout this process. In the first months, there were significant variations in prices. Balloonists, during that period, experimented with their potential audience and tried to determine how high, or low, to set prices, as well as who the target audience should be. Once this settled down, however, prices standardized. In France, tickets cost about three livres or less while in Britain tickets ranged from one shilling to 2*s* 6*d*. These prices enabled a broad range of people to indulge their interest in ballooning. Individuals from the middling sorts on up through the nobility had the economic means to buy tickets. Thus, the willingness and ability of everyday people to participate in this new scientific activity meshed with aeronauts' interest in developing a market for balloons. Through some trial and error, balloonists learned how to advertise this new product and entice new customers. By combining entertainment with Enlightenment, aeronauts forged a new marketplace for the ephemeral and the sublime.[115]

6 BALLOONS INSPIRING CONSUMPTION

The participation of the general public in funding launches, either through subscriptions or ticket sales, represents one of the major ways in which the intro duction of aeronautics coincided with a developing commercial system.[1] While there is a clear connection between launches and increasing active consumer behaviour, this period also sees entrepreneurs taking advantage of the popularity of this new trend to incorporate aeronautics into a variety of other cultural venues in addition to launches. In a growing consumer culture, savants and amateurs alike saw multiple opportunities for selling balloons and balloon-related merchandise to members of the general population. In this instance, rather than developing a new branch of consumer behaviour – namely, trying to attract customers to attend, and pay for, launches – people attached aeronautic ideas to existing areas of the marketplace. Balloons become a theme that crossed through different kinds of material, and immaterial, consumption. For a limited time, lighter-than-air craft made food more exotic, clothes more fashionable, furniture more desirable, plays more exciting and so on. The range of items to which people might apply the new area of aeronautics appeared unlimited expect by their imaginations. This does not mean that every area of balloon marketing achieved great success; but it does imply that marketers seemed willing to apply this trendy new scientific invention to just about any product. As a result, within print culture balloons appear in plays, music, poems, novels, pamphlets, almanacs and treatises. Within material cultural more generally, miniature balloons are sold as toys but they are also appropriated into fashion, images, objets d'art, food and many other areas.

This chapter examines the myriad ways in which entrepreneurs appropriated balloons into different areas of culture. An exploration of the types of balloon-related cultural commodities available will unveil the range of people who had access to this new facet of the economic system and the breadth of possibilities open to those interested in participating. In some areas, such as purchasing a work of art, the field of potential consumers was fairly limited while in other cases, such as access to a poem printed in the newspaper, more people might have access to it. Thus, this chapter provides an analysis of the geography of commercial potentialities available to the public. It is more difficult, if not impossible, to examine directly what people thought about the items they purchased. In the case of these goods, largely unnecessary except as a means of participating in the fashion of the day, people had to imagine a desire to own such objects, go out and purchase them, and then do something with their purchases. They could use them, sell them again, give them away, dispose of them, etc.[2] Very little information exists for the aftermath of the moment of purchase, at least in the case of balloon-related goods. Thus, this chapter focuses on the potentiality of consumer behaviour but does not explore what people may have done with the items once they owned them.

Some people noted the vast potential for balloon merchandise quickly after the invention of aeronautics. An anonymous commentator in the *Freeman's Journal* in 1785, for example, discussed rival ascensions taking place in London and noted that such events 'might bring more people to this metropolis, and cause more money to be circulated'.[3] This author describes the possibilities for ballooning in much the same way municipal officials do today when they vie for major sporting events, such as the Olympics, because of the financial fallout for the area. In any case, it became 'fashionable to speak of balloons' at every level of society. According to one observer, 'the taste for Air Balloon matters had grown to such an extravagant pitch that nothing can pretend to have any intrinsick value in it, unless it has this name as an appendage'. As such, this author claimed even to have heard someone selling 'fine balloon string beans'.[4] Of course, people needed to do more than just speak about balloons; they had to put their money where their mouths were and merchants had to provide objects or services for purchase, which they did. Indeed, 'the taste of the public for aerostatic balloons' quickly 'began to pass into fashion'.[5]

Material Culture

While the theme of balloons appeared in a wide variety of venues, one of the first, and perhaps one of the most obvious, objects for sale were balloons themselves.[6] This is true of larger globes, but in this case the emphasis falls mostly on smaller balloons. Merchants intended these balloons to function fully; in other words, they meant them to fly. A scientific popularizer named Guyot jumped on this bandwagon early.[7] He authored a short essay on the construction and steering of balloons in 1784. On the last page he also announced balloons for sale from his home on the rue Faubourg Saint-Martin in Paris. He offered two different styles, one made from thin goldbeater's skin (typically made from part of an animal intestine) and another, larger and more expensive version made from taffeta. Interested individuals could purchase the smaller balloons in an assortment of sizes including nine, twelve, fifteen, eighteenth, twenty-one, twenty-four, thirty,

thirty-six, and forty-two inches. The cost ranged from three livres, ten sous for the smallest to sixty livres for the forty-two inch globe. The size, and the cost, jumped dramatically with the taffeta version. The smallest of this style was four feet in diameter and cost seventy-two livres while the largest globe, a full twelve feet in diameter, cost 800 livres. Guyot also sold taffeta balloons with diameters of four and a half feet, five feet, five and a half feet, six feet, seven feet, eight feet, and ten feet. If someone wanted to purchase the apparatus for filling the machines, that cost extra. Guyot sold small, medium and large sets of equipment at a cost of twelve, fifteen or eighteen livres. For the more scientifically-minded customer, Guyot also advertised an electrical machine, designed to test the electricity in the clouds, for a mere twelve livres.[8]

Other purveyors of balloons also appeared in France during the first few years. In September 1783 the *Courier d'Avignon* reported that 'novelty merchants had small balloons for sale in their boutiques' and that a lady made a balloon from her taffeta skirt in order to appease the cries of her young son'.[9] An anonymous letter written to the *Journal de Paris* in November 1783 claimed that 'all the world wants to repeat the beautiful experiment of the Montgolfier brothers' and the 'workers of Paris cannot meet the demands of the provinces' for balloons.[10] Even as early as 1783, Jean-Claude Pingeron claimed people could purchase balloons, 'which one at a minimum call scientific'.[11] The same possibility appeared in the *Almanach des Ballons*, a short-lived specialty almanac based in Paris. In 1784 the editors included a short notice about the possibility of procuring globes of various sizes for prices ranging from six livres down to forty sous. The manufacturers of these balloons, two scientific popularizers named Nozeda and Blondy, used a 'material so light, that the globes of thirty inches in diameter did not weigh more than an ounce'.[12] Nozeda was an optician as well as an instrument seller and demonstrator who worked for a time in the Palais Royal in Paris. Less is known about Blondy outside of his activities with balloons; however, he had started advertising globes for sale as early as September 1783.[13] A Monsieur Lebrun, who claimed to have been the official balloonist for the royal children, advertised balloons, along with equipment for the inflation of the globes, for sale in 1791. Interested customers could visit his shop near the Sorbonne or he was willing to ship to the provinces.[14] In one amusing, although probably apocryphal, story a Parisian merchant crated and sent fifty small balloons to a customer in Brussels for use in 'an apartment or a garden'. An official there, however, insisted on opening the case in search of contraband. The box was opened and the balloons sailed out 'to the great astonishment of the viewers'.[15]

British merchants also got into the act. In September 1784 a notice appeared in the *Morning Herald* advertising 'Cheap Air Balloons'. These globes, of 'a new construction', could 'fully answer every experiment that has been, or can be, made by those of a more elaborate and expensive plan'. In addition, 'any person' could launch them no matter what skills you might have; later, however, the notice suggests that these balloons were 'for the convenience of Gentlemen, &c. (who may wish to satisfy their curiosity at a trifling expense, and afford themselves an instructive and rational amusement)'. The size and cost ranged from eight feet, four inches in circumference for two *2s 6d* to a globe seventeen feet in circumference for eight shillings with several additional options in between. Larger sizes were available upon request. The seller also claimed 'the above Balloons are greatly admired for the neatness of construction, and exactness of proportion'. They came with a set of instructions and would be shipped anywhere in Great Britain. Rather hopefully, the merchant also suggested 'Colleges, Academies, &c. may be furnished with any quantity, with written directions for launching them' and that 'the Constructor will wait personally upon any one in or near town'.[16]

Several other individuals also offered balloons for sale to the British public. Allen Keegan, 'at his Oiled Silk and Lawn Manufactory', advertised to 'the Nobility, Gentry, and others, that he has found out a speedy and effectual method of making Air Balloons'. Those interested 'may have them made at a short notice to any diameter, warranted air proof, and at a less expense considerably than any hitherto done in this kingdom'. Keegan, who clearly targeted a fairly elite clientele, also noted some of his other items for sale, including oiled silk, lawn umbrellas and garments.[17] An umbrella maker named J. Clemson offered balloons for sale. He claimed he had a perfect method for 'making Air Balloons to any diameter, on the shortest notice' and at 'most reduced prices'. In addition, the purchaser could have their new balloon filled and 'ready for immediate experiment' or 'left floating at their own houses'.[18] A tin manufacturer named A. Hensley in Cheapside published an advertisement in 1784 in which he sought 'to inform his friends and the public in general, that he has completed a large assortment of his new kitchen furniture', as well as dressing cases, spoons, shower baths and, for two shillings, air balloons.[19]

Arnold Finchett, who also ran a tin manufactory, tried his hand at selling balloons; he published numerous advertisements appearing in August and September of 1784. Finchett claimed his balloons 'have been very much admired for their simplicity, and easy manner of filling with rarefied air only, without any ill smell and the small change of filling them'. His balloons were 'equally visible either by day or night, but more particularly at night' and 'they have been known to travel eight miles from the place where they were let off in the space of ten minutes'. A globe nine feet in circumference cost three shillings. In addition, customers received 'proper directions' on how to use them. Finchett must have faced some competition since he noted that people should 'beware of counterfeits, as an improper sort is vending about town, which do not answer the purpose'.[20] Business must have been fairly good; at the end of September 1784

A ORLEANS CHEZ LETOURMI
L'homme Aux balons Ou La folie Du jour

Figure 6.1

Finchett began offering an expanded line of balloons 'from 9 to 25 feet circumference; and from 1 s. 3d. to One Guinea each'.[21]

In addition to purchasing balloons, real enthusiasts could obtain pictures of balloonists. A miniature engraving of Lunardi sold for half a guinea in London.[22] Lots of other engravings were available in France including images of Jacques-Alexandre-César Charles and Jean-François Pilâtre de Rozier.[23] Jean-Baptiste-Guillaume Curtius's wax museum included a figure of Jean-Pierre Blanchard among its exhibits of famous people.[24] An engraving of the Montgolfier brothers was particularly popular.[25] It was also possible to pay to view the paraphernalia of aeronautic experiments. Thus, John Sheldon threw his hat from a balloon during one ascension and, for a penny, people could see it at the Covent Garden Market. Similarly, an exhibition at King Street included a chance to see 'The Great Balloon, Three Little Balloons, a Pigeon, an Italian Lap-dog, and Monsieur Blanchard', all for a mere one shilling.[26]

Balloons also inspired clothing and other items.[27] The French, of course, dabbled in aerostatic fashions from an early point. In one illustration, subtitled 'the madness of the day', a man wears a balloon shirt, balloon epaulets, balloon shoe buckles, has balloon buttons on his coat and balloon designs on his trousers (see Figure 6.1). A merchant invented balloon hats (les chapeaux au globe volant) as well as 'ribbons à la Montgolfier'.[28] In Boston, Nathaniel Fellows advertised a collection of the 'most superb Balloon Caps & Hatts [sic]' along with many other assorted imported goods; Samuel Barrett's shop also included 'Air Balloon Hats,' in his case coming in white and black.[29] They were still available for purchase in 1802 when a merchant advertised hats 'in the form of a half balloon' based, the author claimed, on 'Parisian fashions'.[30] These hats were intended for women of all social classes, although at least one commentator regretted that trend. 'It is truly lamentable,' a woman wrote to the editors of the *Norwich Packet*, 'to observe what numbers of our sprightly youthful ladies, and our aged dames, at this present time, render themselves ridiculous and contemptible, by aping their superiors' by wearing 'the gay ornamental Balloon Bonnet'. It is so common, she claimed, that balloon hats could be seen on the heads of milliners, dress makers, and washer-women.[31]

In addition to portraits, Lunardi also inspired a variety of other merchandise. As one newspaper noted:

> The Ladies of the town have honored Mr. Lunardi with many marks of their affection. They adorn their heads with Lunardi's hat; they have got the Lunardi colour, Lunardi snuffers, and kitten muffs, in honor of that enterprising puss. A march has been composed called Lunardi's march; and there is an allemande, called Lunardi's flight.[32]

Obviously, Lunardi motivated a plethora of items for sale. When he was presented at the British court it was reported that 'modest ladies appeared in Lunardi bonnets and less modest ones in Lunardi garters'.[33] Balloon hats appeared in England as early as March 1784. A commentator in the *New Spectator* described such a hat as 'considerably encreased [*sic*] in its dimensions, and the decorations over the crown being spheriodical [*sic*]'.[34] One vender, Hartshorn and Dyde's of Wigmore Street in London and on the Circus in Bath, sold 'Air Balloon Hats, either trimmed or plain, or the wires to make them'. These came in 'the greatest variety of colours'. They also advertised balloon fans.[35] With shops in London and Bath, Hartshorn and Dyde's was likely aiming at a fairly well-to-do clientele. Indeed, the *Bath Chronicle* also mentioned the sale of balloon hats in June 1784. Apparently, such hats suffered a brief fall from grace as the fashion item of the season when, for a few days, 'the rural straw umbrella' took over. However, they recovered their place of honour and have 'again been re-inflated as the capital ornament of female undress'.[36] These hats also reached other parts of the globe; apparently, New Yorkers fell victim to the balloon hat during the summer of 1784 thanks to some inspiration from Lunardi.[37] A 'Balloon Song' claimed that 'the bonnet, hat, cap and tupee' as well as petticoats, were fashioned after balloons.[38]

Along with hats, women also could have their hair styled in a balloon-like fashion. The same 'Balloon Song' mentioned above wrote that 'Ev'ry lovely charming she / Must have her hair – Balloon O'.[39] In Paris, women had their hair styled after globes while the men purchased shoe buckles with balloons. Men could also purchase waistcoats with balloons embroidered on the pockets, bags and cases with aeronautic designs, and miniature balloon buttons. There were canes with a balloon head and swords with aerostatic designs.[40] So much fashion was devoted to balloons that the *Affiches de Toulouse* suggested that 'to follow the torrent', they might 'given an entire issue over to globes'.[41] The arrival in London of the French hairdresser Sieur Dugrenier caused some notice. Dugrenier had trained with Leonard, the personal hairdresser to Queen Marie Antoinette who valued him so highly she even took him with her when she and Louis XVI tried to flee France in 1791. Dugrenier proposed to style hair in 'all the modern genres', including after balloons.[42] Antoine de Rivarol fretted over the popularity of aeronautics. In addition to complaining of people who formed societies about balloons and who included them in theatrical productions, he also mentioned that 'women dress their hair like globes'. This enthusiasm, he thought, caused some astonishment among foreigners in France.[43]

Few areas escaped the fashion for balloons. Ear-rings, worn by Parisian ladies, appeared with air balloons as the pendant.[44] In the United States, a French lady who had just arrived reported the theft of a number of objects, including 'one pair of Gold Ear-rings, in the form of a balloon'.[45] A certain Mr Hills, the

'Glass-Stainer to his Royale Highness, the Prince of Wales' created 'Air Balloons in Stained Glass' which he thought would 'perpetuate the memory of that great event for ages to come'.[46] The same merchant also developed a balloon-shaped pincushion.[47] One commentator claimed a 'balloon influenza' raged with 'balloon hats, balloon bonnets, balloon caps, balloon ribbons, and balloon pins' as well as 'double balloon ear-rings, and balloon side curls so that there is not less than seven balloon articles appertaining to the decoration of the most beautiful balloon in nature – the head of a pretty woman!'[48] Dresses were designed 'au ballon', which might leave the chest area 'artificially and artistically inflated'.[49] Someone in Paris developed a balloon game, advertised as a 'new aerostatic game to be used by elevated spirits'. Players moved their tokens from area to area each representing a well-known flight that had taken place between 1783 and 1785.[50] Even books might be bound with a balloon design on the cover.[51]

Food also came under aerostatic influence. Apparently, balloon oysters were for sale in New York City while balloon-themed gingerbread was sold in London and New York.[52] A restaurant in Paris sold a 'filet à la montgolfier'.[53] John Greening, a merchant in London sold a 'Balloon Sauce for Fish' and 'dishes of all kinds, which has met the approbation of his friends in general' for over two years. Bottles sold at eight shillings for the largest to two shillings and three pence for the smallest.[54] In Paris an enterprising merchant even created a new liquor called 'crème aérienne' or 'crème aérostatique'.[55] In Charleston, South Carolina a type of Madeira was sold 'by the title of Air Balloon' at the shop of James Miller.[56]

Balloon-themed furniture appeared on the market although these objects were clearly designed for a wealthier clientele.[57] A commode, with an inlaid design representing the first hydrogen flight performed by Jacques-Alexandre-César Charles, appeared along with chairs, console tables and trunks. People also ordered glassware shaped like balloons as well as tapestries and wallpaper made by the Réveillon factory in France.[58] Balloons provided the inspiration for new chandeliers as well as snuff boxes and clocks.[59] A cavalry officer named Doray even presented Louis XVI with a 'clock of his own invention' which looked like a Montgolfier balloon. Doray designed it in such a way that on the last stroke of twelve on Sundays, a gondola rose up 'in which are seen figures representing two aerial travelers'.[60] In 1808 a man from West India moving to London sold off a good deal of his belongings including an 'elegant balloon clock, said to have cost 500 livres'.[61] A large amount of porcelain and ceramic objects appeared with balloon images on them.[62] These include, of course, plates and cups but also pitchers, sugar bowls, mustard pots and, somewhat less exaltedly, chamber pots.[63] Earthenware tiles were also produced on which artists painted images of particular balloonists, in their globes.[64] More expensive balloons appeared in the artwork of the day as well as in monuments, many of which were devoted to aeronautic exploits.[65]

A wide variety of songs also utilized the balloon theme as did, in at least one instance, a full-scale opera.[66] Some of these songs were written in the local patois. These include some from Lyon, for example, including: a 'Chanson du ballon de M. de Montgolfier' which praised the Montgolfiers and Pilâtre de Rozier.[67] Another Lyonnais song arrived on the scene in 1784, this time celebrating the 'balloon des Brotteaux' and sung to the tune of a Christmas carol.[68] A popular song from a play performed in England included the lyrics 'Your coaches and chariots henceforth lay aside, / Prepare in Balloons through the skies to ride'. But instead of relating this new invention to the science of the day, the song went on to add that aeronauts were 'like witches on brooms', a curious analogy at a time when ballooning still reigned as a popular scientific discovery.[69] Other songs were more traditional in lauding balloons. A song, called simply 'Air Balloon,' included the lyrics 'Since ev'ry man now builds his castle in air / So high, you can't see it at noon: / A curious invention is found to get there, / To fly in an air balloon'.[70] While individual aeronauts and ballooning in general served as inspiration for songs, some of them also focused on other aspects including, for example, the sheep that went aloft from Versailles in 1783. In a song titled 'The Aerial Sheep', the lyricist wrote how the sheep was 'cherished by mortals and the gods'.[71] Boucault's song, played to the air of 'Où allez vous Mr. l'Abbé', offers a musical critique of Abbé Miolan and his partner Janinet. Their 'failure to launch' led to a riot and the destruction of the balloon leaving Boucault to ask 'who are the kings of the airs?'[72] Another song on this topic, this one played to the air of 'Du vaudeville de la Rosiere', referred to 'Miolan the ignominious' and his 'useless flight'.[73] In 1818 students wishing to purchase sheet music for the flute could buy a song entitled 'Puss in the Balloon' from Riley's Flute Melodies catalogue for a mere fifty cents.[74]

Literature

> Inevitably these accounts make somewhat repetitive and therefore monotonous reading, the more so because literary ability and the desire to ride in a free balloon seldom seem to go together.[75]

While balloons inspired a whole slew of merchandising, arguably printed works became the most prevalent location for balloons in the culture of the period. Much has been written about the power of print in early modern Europe, and in eighteenth-century France in particular. Historians, including Roger Chartier and Robert Darnton, have elaborated on the manner in which ideas moved, shifted, and were appropriated by different groups through the vibrant print culture during the period of the Enlightenment and the French Revolution.[76] Balloons entered into this cultural milieu with a vengeance and were quickly utilized in a vast array of printed forms including poems, plays, prose literature, almanacs, scientific treatises, travel accounts and in the periodical press. Literary interest in aeronautical matters continued a long tradition of using science as a theme in poetry and prose.[77] Peter Gay noted, in his history of the Enlightenment, how many poets lauded Isaac Newton. He also suggested, however, how 'from country to country and decade to decade, the tributes to Newton changed little. They were always fervent, usually sincere, but in the long run mechanical and monotonous', a point of view echoed by L. T. C. Rolt in the quotation at the head of this section. Gay forgave the poets their monotony noting they 'claimed neither to solve scientific discoveries nor to provide scientific information; their task was to reflect and diffuse a new attitude toward nature, toward knowledge, toward the world'.[78] This section will not provide a literary analysis of balloon-themed literature; instead, the focus will be on the range of ways in which balloons were used by authors.

Contemporary writers certainly picked up balloons early and eagerly, with poets producing verse in vast quantities. Even at the time people noted how much poetry, and how much bad poetry, appeared thanks to the new globes. Most of this poetry appeared in the vernacular although a few tried to raise the bar and produced works in Latin.[79] One anonymous commentator noted it was the 'unhappy destiny' of Jean-François Pilâtre de Rozier to have inspired in poets 'bad verse' despite the innate excitement balloons engendered in people.[80] Admittedly, a few well-known and famous poets penned balloon poetry. The young Percy Shelley, for example, wrote a sonnet 'To a balloon, laden with Knowledge'. He writes that the balloon will 'fade like a meteor in surrounding gloom' but the fire carried on board, perhaps the 'knowledge' of the title, will remain 'A beacon in the darkness of the Earth, / A Sun which o'er the renovated scene / Shall dart like Truth where Falsehood yet has been'.[81] Anna Letitia Barbauld also wrote a poem, 'Washing-Day,' in which children, including the young Barbauld, watch their mother do the laundry while they blew bubbles 'little dreaming then / To see, Montgolfier, thy silken ball / Ride buoyant through the clouds – so near approach / The sports of children and the toils of men'. Here, children imagine the possibility of aeronautics even before their invention.[82] Erasmus Darwin made reference to ballooning, in particular the death of Pilâtre de Rozier, in his 'Economy of Vegetation'. Somewhat tritely, he refers to Pilâtre de Rozier as the 'hapless Icarus' as he 'rushes through the affrighted air / With limbs distorted, and dishevel'd hair, / Whirls round and round, the flying crowd alarms, / And Death receives him in his sable arms!'[83]

Most of the poetry printed, however, appeared without attribution. These poems can be fit, roughly, into several categories. First, many of these poems laud specific aeronauts. Often, these poems were quite flattering in their praise if not downright obsequious. A 'Constant Reader' named Emma in Birmingham, for example, sent a short poem in to the local newspaper after Mr Harper's ascension in that city writing 'Hail then, distinguished Youth! Britannia's boast, / Apollo's darling, and the reigning toast'.[84] Richard Crosbie received similar attention although the anonymous poet in this case acknowledged that some people vilified Crosbie's name out of 'envy'. The author wrote, however, 'the tribute due to great attempts is thine / You, of Hibernia's Sons, none can deny, / A Dedalus, first launch'd into the sky'. The poem goes on to compare Crosbie to Christopher Columbus.[85] Another author describes aeronauts as 'the new Argonauts' and, in addition to Columbus, compares Montgolfier with Captain James Cook.[86] Jacques-Alexandre-César Charles and Robert are given high honours in many poems; in one such effort 'they have marked their place / between men and the gods'.[87] Naturally, the Montgolfier brothers received their share of these missives: 'Montgolfier, who all of Europe / could not revere enough'.[88] The sudden death of a balloonist also prompted poems. James Sadler's demise, when he was thrown from his balloon, led to the publication of a broadside with 'Lines on the Lamented Death of Mr. Sadler,' where the poet wrote 'The Hero rode majestic through the air, / From cloud to cloud, not thinking death was near'. Later in the poem 'Death shot his shaft, his rude, relentless dart, / And pierc'd the short liv'd victim to the heart'.[89]

Vincenzo Lunardi received considerable poetical attention throughout his aeronautic career. In one case he is compared to Jason and the Argonauts in a short poem appearing in the *Morning Post* in 1784: 'Britons, that vanquish even Greeks in arts! / To see one daring Modern venture more / Than ever kings or heroes had before'.[90] In an ode from 1790 he is 'the flying Lunardi' who 'aspires to the airs'.[91] More prosaically, the author of a London street ballad in 1784 said 'The sky being clear a fine sight we did obtain / Success to Lunardi and his air Balloon'.[92] After his first flight in England, Lunardi inspired a poem in which the author wrote 'Thus brave Lunardi dar'd a space unknown, / And Fame bestows him an immortal crown'.[93] Another poet described Lunardi after this event as 'The daring youth who claims th' admiring song!'[94] Just as Crosbie had been compared to Columbus, Lunardi also received such recognition in a work by Manoel Maria de Barbosa du Bocage where he likened Lunardi to Vasco de Gama, Columbus, and Magellan.[95] James Tytler, the Scottish aeronaut whose experiments never quite took off, wrote a lengthy poem in honour of Lunardi's visit to Scotland; Lunardi, rather immodestly, included the work in his book describing his Scottish experiments. In the poem, Tytler greets Lunardi with 'Welcome to earth, to feast our longing eyes!' and goes on to recount his own misfortunes compared to Lunardi's successes.[96]

Jean-Pierre Blanchard also received considerable attention from the poets. When he crossed the English Channel with John Jeffries, for example, a short work appeared in which Neptune rears up and complains 'Britannia to me shall no more pay her vows'.[97] In somewhat typical fashion, Blanchard even composes some verses of his own in which he writes that 'wise and modest / Montgolfier taught us the air of crossing the winds' but, he goes on to add, Blanchard 'will do the rest'.[98] In 1784 Marie-Emile-Guillaume Duchosal wrote a poem, in two parts, lauding Blanchard and his launches. Duchosal later doubled the size of the poem.[99]

While many individuals received attention from poets, often works simply described aeronautics in general. An early missive to 'aerial navigators', for example, asked the typical question seen in poems, 'Are you mortals or demi-gods?'[100] A poem in the British newspaper the *Morning Herald and Daily Advertiser*, exhorted would-be aeronauts to hurry up, get busy, and catch up with the rest of Europe. The poet asked, 'Shall Albion's sons for Fame, for arts, renown'd, / Consent to crawl like reptiles on the ground?'[101] A poem by the French author Arnaud de Saint-Maurice described aeronauts as 'the children of Newton' and claimed 'A thousand savants have already, with a glorious ardor / Worked day and night on the Famous globe'.[102] Some of these poems were quite lengthy. Vincenzo Lancetti, for example, published his *Areostiade: Ossia, Il Montgolfiero* in two volumes due to its size.[103] In 1786 Thomas Seddon's bookstore in Philadelphia advertised a poem called simply *The Balloon* that sold for 1*s* 10*d* while William Woodhouse's bookstore sold the poem *Balloon: Or an Aerial Trip in the Celestial Regions, A Poem in Six Parts* for a quarter of a dollar.[104]

The wondrous nature of ballooning often appears in these poems. An anonymous author in England wrote about the novelties found in Paris and exclaimed 'such surpize as the air-balloon, lately seen there, / Which men, pigs, and chickens, conveys thro' the air'.[105] After the invention of parachutes, they also came under scrutiny in poems. 'In London', claimed one poet, 'there's something always to behold / To please Johnny Bull and to pocket his gold. / So balloons are reviv'd by a Frenchified lad, / And all the world surely are now balloon mad'. This 'lad' was André-Jacques Garnerin who 'went up, which increas'd his repute, / And came safe to the earth in his grand Parachute'.[106] Another poet claimed, satirically, that trade 'shall profit by this airy navigation', and that 'Smart milliners shall crowd the air balloon, / To bring new fashions weekly from the moon'.[107] This sort of hyperbole was not uncommon. In another work the author claimed 'In my chariot aerial, how pleasant to go, / To see all my friends in the stars: / Take a breakfast with Merc'ry and dine if I please / With Jupiter, Saturn, or Mars!'[108]

Balloons also offered poets a chance for various sorts of criticism. An American poet described a failed launch in terms of the Federalist debate. After the shamed aeronaut returns to land he says:

> 'Your cause wants virtue more refin'd
> Before admittance you can find
> To higher regions; federal fire
> Can but consume, nor raise us higher.
> Now be convinc'd, by what you see,
> That men were made for liberty'.
> Then sudden bursting to a flame,
> It quickly did consume the same;
> So good by, now, to Adams John
> And Louis Tousaint's Air Balloon.[109]

The unsuccessful ascension is linked here to the Federalist debate over the structure of the new American government system and the relationship between the central government and the states. William Pitt, the younger, is parodied in a poem, titled 'Pitt's Balloon', in which he is seen in association with the Devil.[110] The battle between Sadler and Lunardi for primacy in England resulted in a short poem in which the author writes 'Behold a windy competition / Two puff-makes in opposition, / The whole must end in vapour'.[111]

In some cases, poems made a reference to the interest of women in ballooning. One short work, 'Lines on a Young Lady wishing to ascend in an Air Balloon', lauds the beauty of a potential aeronaut.

> Forebear, sweet Girl, your scheme forego,
> And thus our anxious troubles end;
> That you will mount full well we know,
> But greatly fear – you'll not descend!
> When angels see a mortal rise,
> So beautiful, divine, and fair;
> They'll not release you from the skies,
> But keep their sister angel there![112]

Whether the poet wanted to bestow flattery on the young girl or had sincere worries about her not surviving the excursion remains unclear. Some of these poems referred to balloon fashion and, often the number of women who seemed to find Vincenzo Lunardi attractive. Thus, after one of his ascensions inspired a vogue in 'Lunardi garters' an anonymous poet claimed 'Below her knee, or round her thigh, / His dear enchanting name did tie; / And shew'd the bold advent'rer more / Of *Heaven* than e'er he saw before'.[113] The implication seems clear; there were some women so enamoured of ballooning and Lunardi that they allowed him to see up their skirts as far as 'heaven'. This theme appeared elsewhere. A short 'Caution to those Ladies who have ventured to wear Garters inscribed with the name of that rising Genius Mr. Lunardi,' suggested:

> Ye bright British fair,
> Who love bubbles of air,

And for fashion your safety will barter;
Lunardi's the bold
At your knee can't be cold,
He'll rise to the *Zone* from the Garter.

Here, the 'zone' probably replicates 'heaven' in the earlier poem and certainly implies people perceived Lunardi as something of a libertine (if only because he was Italian), who used his aeronautic adventures to woo women. A popular engraving at the time, titled 'Love in a Balloon', shows a man and woman embracing in a balloon as it ascends. The man says 'Ah Madame, it rises majestically,' to which the woman replies 'I feel it does Signor'.[114]

In addition to their quick and continuous appearance in poems of all shapes and sizes, balloons also landed fairly rapidly in the middle of plays. Playwrights took advantage of this new invention to move their characters around great distances in relatively short amounts of time. Thus, balloons were less of a central motif in drama than they were a plot point that allowed the main characters to travel to a foreign land, or perhaps someplace even further distant such as the moon, and escape their troubles or interact with other people. Elizabeth Inchbald's play, *The Mogul Tale, or the Descent of the Balloon*, offers one example. Here, Johnny, a cobbler, Fanny, his wife, and a Doctor launch themselves in a balloon only to get lost and end up in the fictional land of a Mogul. The locals are amazed by the descent of the balloon, which they initially mistake for a 'great ravenous bird' or a 'chariot of some of the gods'.[115] This description mirrored the manner in which aeronauts reported peasants responded to their arrival in their fields and villages. The aeronaut here is the Doctor, who is not a particularly good balloonist. He cannot control the globe: as Johnny complains, 'the Doctor, with all his magic, could not stop it [the balloon] when it was set a-going'.[116] The Doctor might represent Lunardi in this play. Fanny confronts him at one point claiming he is a bad man who wanted to kiss her during the voyage when Johnny was asleep.[117] Later, the Doctor has to defend himself to the Mogul and define balloons. He tells the Mogul balloons are 'a Machine of French invention, founded on English Philosophy,' which had brought them to the Mogul's land against their will 'without any intention to seduce away any of the females of the Seraglio.[118] Inchbald's play was adapted elsewhere and appeared in New York City as *The New York Air Balloon*, and in Boston as *The Boston Balloon; or, a Supposed Flight from the Beacon-Hill*.[119]

Another British play, Frederick Pilon's *Aerostation, or The Templar's Stratagem*, focused more directly on aeronautics although, in this case, his intent was to ridicule the rage for balloons.[120] One of the characters, a widow named Mrs Grampus, has been overcome with balloon mania. She greets the local lord, the Baron Bubblebergen who wants to ascend in a globe, for example, while wearing her 'balloon hat', 'balloon sack' and 'balloon apron'.[121] The Baron is an imposter, as it turns out, who is trying to turn her attention away from her current lover, a bookseller named Quarto who has promised to go up in a balloon for her. In the end, Quarto and the false Baron almost have a duel, Quarto's nephew ends up marrying his love, Sophia, and everyone is, more or less, happy. The main goal of the play, however, was to 'ridicule the use of balloons' and, according to one review, this goal 'was amply fulfilled'.[122] Other plays also appeared in Britain with balloon themes, or sometimes actual balloons. At Astley's Amphitheatre a play was performed in 1784 with the title *The Air Balloon*. In 1785 another piece, *The Cobbler and the Air Balloon*, appeared at Astley's, whose goal was to show 'there was some danger in Aerostatic Experiments'.[123] At the Drury Lane Theatre a play appeared in December 1783, called *Fortunatus*, 'in which will be introduced an Air Balloon'.[124] It is unclear if the balloon had anything to do with the play.

French playwrights also turned their attention to balloons. Nicolas Marie Félix Bodard de Tezay wrote *Le Ballon, ou La Physicomanie*, in which he poked fun at the mania for aeronautics.[125] This appeared in 1783. He followed this the next year with a play called *Arlequin, roi dans la lune*, which used the balloon as a device to get his characters to the moon although the balloon itself was not particularly central to the play much in the same way as Inchbald's *Mogul's Tale*.[126] Bodard de Tezay's plays, along with others such as L.-A. Beffory de Reigny's *Nicodème dans la lune, ou, la Revolution pacifique*, did have radical overtones in that he used his geographic distance, in this case on the moon, to criticize the social order.[127] There were also two French plays which used the parachute to move their characters around. In Dominique Boutard's *La femme en parachute*, Victoire parachutes down to save her marriage.[128] In another play, *Le Parachute*, one of the characters, a physicist named Gilles, woos the fair Colombine at her window in a balloon and, eventually, invents a parachute and wins her over.[129] In M. de la Montagne's comedy, *La Physicienne*, the lead female character, Madame Siphon, even dares to send to the scientific academy in St Petersburg a 'discourse on the theory of aerostatic machines'.[130]

As with plays, the appearance of balloons in novels often centred on their use as a tool to get the main characters moving. In the case of *The Balloon, or The Aerostatic Spy*, for example, the hero travels across the world and provides social commentary on what he sees.[131] The author purports to meet an old gentleman one day when returning to London from Dover after watching one of Blanchard's ascensions. The old man 'had in his possession the papers of another gentleman, whose affairs had called him to Ireland'. Our author convinces his companion to let him publish the papers 'for the Instruction and Amusement of our Countrymen'.[132] The papers tell the story of a young man, born in America, tutored by his alchemist father. After his father dies, he escapes an unwanted marriage to his homely cousin, meets a beautiful girl, named Lucy, flees to the sea, is attacked by pirates and shipwrecked, then meets an old man named Sagely who tells him about balloons. They eventually make one from animal skins and the young man goes to explore Africa, Constantinople, Indostan and then Lisbon. Later stops include the Pyrenees, Spain, Paris and England. In England he lands in Maidstone, London, Windsor and Bristol. He eventually meets up with his beloved Lucy whom he marries. A number of other, similar, books also appeared including *An Account of Count d'Artois and his Friends' Passage to the Moon, in a Flying Machine, called, An Air Balloon.*[133] The anonymous novel *Aventures singulières, d'un voyageur aérien* follows the story of a aeronaut who uses his balloon to travel around the world with a love story as the subplot.[134]

Other books appeared throughout Europe. Jean-Paul Richter published his fake balloon account, *Des Luftschiffers Giannozzo Seebuch*, in 1801.[135] Several books appeared fairly quickly in French, published in various places in France but also in Brussels, including *L'Amour dans le Globe*, which was a collection of poems, prose essays and other pieces; *Le Mouton, le Canard, et le Coq*, a dialogue between the first three animals to go up in a balloon; *Réclamation du Mouton, premier navigateur aërien*, a silly look at the sheep on that first flight; and the *Histoire intéressante d'un nouveau voyage à la lune*, another story of someone going to the moon in a globe although in this case he returns to Paris with a very pretty lunar lady.[136] François Suleau's *Voyage en l'air*, published during the early years of the French Revolution, recounts another trip to the moon, using one of Blanchard's balloons. Once there, the aeronaut encounters an orator who speaks out against the local prince; eventually, the author becomes an advisor to the prince and is adopted as a citizen.[137]

Balloons also appear in other ways in works of fiction. Claude Bretin's *Contes en vers*, for example, includes a tale about a balloon and its use in a love triangle. Eugenie is married to Pedro, but loves Gusman. Pedro locks her in a high tower but Gusman uses a balloon to stop by and woo her.[138] In another love story, Amour is going to help the hero Dorval find his true love Alida; to do so, Amour uses a 'balloon chariot' hitched to two dragons who 'know the routes of the Empire'. Dorval ends up searching for Alida in America and they finally meet up in Philadelphia.[139] One of the most famous balloon-related novels was the anonymous *Le Retour de Mon Pauvre Oncle*. This is a short novella in which a man, by virtue of hydrogen, ends up on the moon and describes what he sees there.[140] This novel may have stemmed from a 1783 satire, published in France but reprinted elsewhere, in which a man writes a letter cursing balloons and claiming his uncle, who was obsessed with balloons and wanted to be the first man to ascend in one, flew out of the window after accidentally ingesting some hydrogen.[141]

Balloons, of course, also entered into print culture in a myriad number of smaller ways. Quick references or satirical asides are easy to find, for example, in the periodical press. In 1820, an 'Extract from the Journal of a Balloonist' included descriptions of hail banks, from which the author grabs a hat full of hailstones the size of pigeon's eggs, the marriages taking place on Venus, and the beat of war drums heard from Mars.[142] A fake shipping news report, based on what might happen if balloons replaced ships, describes 'the good balloon Daedalus, Capt. Wingate, will fly in a few days for China; he will stop at the top of the Monument to take in passengers'.[143] A notice to the inhabitants of New Jersey announced a new land speculation scheme for the moon. The notice claimed 'the climate is the most healthy' and there is 'no Indian Savage to murder or scalp you either on your passage nor when you arrive thither'. Those interested were notified 'the Balloon is to set off the first Monday in every month from the top of the Allegany Mountain, at the rate of three dollars for every passenger'.[144] A snub against the Scottish appeared in 1785 when an anonymous commentator noted that Scotland was 'equally behind hand, with Ireland, in balloon philosophy' even though it was a country which needed 'little temptation' to 'induce a man to fly away!'[145] Examples such as these appeared frequently.

Periodicals, Almanacs, Narratives and Manuals

Certainly, the periodical press offered the easiest method of obtaining information about balloons. Newspapers and journals filled their pages with accounts, descriptions, advertisements, poems and many other sundry references, fact-based and fanciful, to aeronautics. In Italy, one publisher printed a newspaper devoted entirely to balloons, the *Giornale aerostatico*, although it only lasted one year.[146] In 1784 and 1785 the *Journal encyclopédique* had a permanent section on 'Diverse notices concerning the aerostatic machine'.[147] For those wishing for more information about this new world of opportunities, other printed material appeared to satisfy their needs. Publishers, for example, created aeronautical almanacs. As almanacs already held an important place in print culture and enjoyed a readership from all levels of society, this proved a likely money-maker. In France, the *Almanach des Balloons, ou Globes Aérostatiques; Etrennes du jour physico-historique et chantantes* was available in 1784 although no other volumes seem to have been printed. In North America, *The Balloon Almanac* was published beginning in 1785 and was still available in 1790 although it is unclear how much it really focused on aeronautics. The 1787 edition included, among others, sections on eclipses and moveable feasts tides, phases of the moon, the values of weights and coins, and the location of Quaker meetings.[148]

Of course, there was nothing like first-hand information describing aeronautical experiments. The earliest experiments came with accompanying published accounts. Jean François Pilâtre de Rozier, the Robert brothers, and J.-A.-C.

Charles, for example, wrote pamphlets outlining their experiences.[149] Jean-Pierre Blanchard almost always wrote letters to the local newspapers describing his voyages and a number of these were published as pamphlets.[150] In some cases, he had these translated into English so he could address a wider audience. While in America he also published, with Gardiner Baker, a brief history of ballooning, written in English.[151] Vincenzo Lunardi followed this same pattern, to some extent. He wrote descriptions of his travels which were then published in a number of newspapers. But he also wrote two books, collections of these stories with additional letters from members of the audience.[152] Lunardi composed these accounts, he claims, out of 'infinite concern' having observed 'that some narrow-minded men are attempting to avail themselves of the curiosity and benevolence of the public, in a manner which, he hopes, every generous mind will revolt at'.[153] In other words, some people were publishing their own versions of his voyages and he wanted to make sure he got the credit. Other aeronauts also published descriptions of their voyages. Thomas Baldwin's lengthy, moment-by-moment report included colour pictures of the landscape that moved beneath him.[154]

A number of savants published histories or studies of ballooning. One of the earliest, and most famous, records in English is Tiberius Cavallo's *The History and Practice of Aerostation*.[155] This remained the standard early history of aeronautics for some time. In France, the balloon enthusiast Barthélemy Faujas de Saint-Fond collected letters, articles and other descriptions about aeronautics and published them in the two-volume *Description des expériences aérostatiques de MM. De Montgolfier*.[156] Other such studies abounded including Thomas Martyn's *Hints of Important Uses, To be Derived from Aerostatic Globes*, Guyot's *Essai sur la construction des ballons aérostatique et sur la maniere de les diriger*; Carlo Amoretti's *Delle Macchine Aerostatiche*, and Pierre Bertholon's *Des avantages que la physique, et les arts qui en dependent, peuvent retirer des globes aërostatiques*.[157] Some of these books were devoted to general explorations of aeronautics while others focused on some particular aspect, such as steering them. A number of these books went through multiple editions. While authors published many complete books on this subject, it might also be the case that a part of a book might be devoted to ballooning. In this way, entries on ballooning were written for encyclopedias and specialized dictionaries including Mathurin Jacques Brisson's *Dictionnaire de la physique* and the American edition of a work entitled *Encyclopedia, or a Dictionary of Arts, Sciences, and Miscellaneous Literature*.[158]

Whether in poems, plays or novels, as furniture, food or clothes, the advent of aeronautics left an indelible mark on the culture of late eighteenth- and early nineteenth-century Europe. People encountered ballooning everywhere they went. Newspapers and journals were replete with references to the new invention and those with money to spare could purchase goods in an astonishing variety of forms, and at a dizzying array of prices, enabling them to bring ballooning from the heavens and into their homes. While launches constantly functioned as a form of entertainment, combined at times with experiment, to which people could go, they also became appropriated into just about every other form of cultural activity available to people at the time.

CONCLUSION: ROMANTICIZING ENLIGHTENED BALLOONS

The popularity of balloons continued long after this study ends. They appeared in paintings and books, including Edgar Allen Poe's short story 'The Balloon Hoax' as well as in Jules Verne's novel *Five Weeks in a Balloon*.[1] They remain a form of popular entertainment today. Tethered balloon rides are fairly common. Interested individuals, for example, can pay for an outing that takes off from the Parc André Citroën in Paris or at Navy Pier in Chicago. Calendars are available for purchase with balloon themes, some focusing on particular festivals. Of course, the festivals also draw crowds, many of whom come to watch the launches or wish to view the balloons during the inflation process. This is especially true for night-time events during which the interior of the balloon might be illuminated. At the same time, attempts to circumnavigate the world in a balloon, undertaken by wealthy adventurers such as Steve Fossett and Richard Branson, occupied the headlines throughout the 1990s and into the twenty-first century. Fossett eventually succeeded in his quest when he travelled around the world in just over thirteen days from 19 June to 3 July 2002. Arguably, then, the continued popularity of aeronautics rests less on their enlightened qualities – such as their potential utility or role in demonstrating humanity's ability to conquer nature – or even because of their appropriation into social and economic life. Instead, balloons undergo something of a transformation during the nineteenth century due to their inability to meet the standards of enlightenment. In this manner, balloons shifted from being a failed example of the Enlightenment to a shining example of Romanticism. Richard Holmes, in *The Age of Wonder: How the Romantic Generation Discovered the Beauty and Terror of Science*, argues that the development of ideas connected with Romantic science, including the idea of a solitary genius and the 'notion of an infinite, mysterious Nature, waiting to be discovered', coalesced in the invention of balloons which had clearly identifiable figures, aeronauts, launching themselves boldly into a part of the natural world, the heavens, to which people had never before ventured.[2] The aeronautic experiment, and even the fiction of the aeronautic experiment, gave way to the balloon adventure complete with hero who battles nature instead of trying to understand it.[3]

Put in these terms, of course, the transition from enlightened to romantic balloons is a false dichotomy; the two aspects of aeronautics always coexisted just as it is arguable that Romanticism grew out of the Enlightenment instead of existing in opposition to it. Balloons help unravel the complex interconnections between these two movements. From the moment of their invention, the emotional fervour with which audience members reacted to launches and idolized those involved clearly speaks to a romantic sensibility. These responses occurred almost from the very start of aeronautics, perhaps in spite of repeated attempts to rationalize ballooning and turn it into something useful. Similarly, experiments using balloons as a rational tool for gaining access to the atmosphere or hard-to-reach locations, like the arctic, continued to take place in the nineteenth century.[4] Balloons were always romantic, and continued to offer opportunities for enlightenment. Even as aeronauts conducted experiments, audience members swooned, gasped and generally emoted belying the assertions of those who wanted to confine the new invention to the realm of science and reason. In this way, balloons epitomize the overlapping nature of the Enlightenment and Romanticism as well as the lack of control individual savants, or even small groups of savants, had over aeronautics. Nobody really had control over how globes would be viewed or understood by the general public.

For these reasons, the most critical moment in the early history of aeronautics came not with their first demonstration of a balloon but the first such event before an audience; in addition, an even more crucial moment took place when a similar exhibition took place before a *paying* audience. Joseph and Etienne Montgolfier's decision to use amateurs as witnesses to their success altered the trajectory of aeronautics in ways they probably did not anticipate or want. Although this particular group of observers gave their launch the necessary gravitas and status needed to convey to the state authorities in Paris and at Versailles the Montgolfiers' sincerity and the truth behind their claims, the utilization of spectators simultaneously placed this invention in the hands of amateurs.[5] Similar benefits, and problems, occurred with other attempts to draw the public into scientific debates. Backers of animal magnetism and dowsing, for example, used affidavits from witnesses to help support their claims.[6] Popular support, however, did not readily translate into scientific acceptance.

When Barthélemy Faujas de Saint-Fond placed the next big launch in the hands of a paying public, things really changed. The success of his endeavour meant the state saw no need to provide money for a science so readily funded by the general public. As a consequence, savants found themselves on the margins of aeronautics while amateurs and entrepreneurs carried the day. Those balloonists willing and able to work within the emerging marketplace had an easier time of it. Rich amateurs notwithstanding, the bulk of ballooning ended up squarely in the public sphere. Savants could conduct aeronautic experiments, but frequently did so with an audience present whether they wanted one or not. In this manner and from its earliest days, ballooning emerged as a form of mass science which depended on the public for support.

Although balloons certainly offered a lot in terms of spectacle and entertainment, it still remains a little unclear why people who hitherto had not invested much of their time and money on science now felt a desire to do so. Certainly, participation in science, through books, public lecture courses, and in other areas, grew during the course of the eighteenth century. Aeronautics, however, expanded this participation far beyond what it had been. A number of theorists have suggested reasons why people buy things. Thorstein Veblen, Georg Simmel, and Norbert Elias and Neil McKendrick, for example, focus on the notion of emulation.[7] In this formulation, individuals would have participated in subscriptions or would have purchased tickets for launches in imitation of people higher up in the social and cultural hierarchy. For example, since the royal court of France witnessed a launch in 1783, others in France would have wanted to do the same, thus leading to subsequent interest in such events. Similarly, the Prince of Wales honoured Vincenzo Lunardi by attending one of his experiments; his presence may have encouraged others to join so they could be seen associated with royalty.[8] Those launches that included a published list of the subscribers lend themselves to the emulation thesis; the prestige that came with having one's name published with the social and cultural elite would certainly have encouraged a certain amount of imitation.[9]

The emulation thesis, however, fails to explain fully the overall breadth of the participation in ballooning activities. In most cases, the subscribers or ticket holders were not identified in a public forum. In addition, at no time were balloons completely in the hands of the elite. As a new commodity, balloons had yet to be socially or culturally defined and, as it turned out, the elites did not automatically become associated with the new practice of aeronautics. On the contrary, instead of being controlled by the court or scientific academies, balloons fell under the dominion of a much broader range of individuals who sought out the financial support of members of the middling class right from the start. Arguably, the elites might have emulated the middling sorts in this case rather than the other way around.[10]

Another suggestion adopted by some historians trying to explain how and why people spend their money concentrates on the relationship between objects and their meanings. In this instance, consumed things operate as signs that the consumer used to help in identity formation or self-fashioning.[11] As Joyce Appleby suggests, 'the study of consumption gives us a window on the elaboration of personal identity. Consumption offers people objects to incorporate into their lives and their presentation of self.'[12] While this explanation may help in the analysis of the consumption of material objects related to balloons, it does less in aiding our understanding of why people purchased tickets for something as transient as a balloon launch. The group experience was certainly critical, and the ability to recall specific memories of a launch and claim attendance might have become an important marker for people, but the evidence for this is largely absent. While being part of the group on hand for an important event might have been important, it would have been difficult to display this to observers in the same way owners of material objects could have displayed a balloon clock, plate or watch. Thus, while it may have been that something like respectability, discussed by Woodruff Smith, necessitated the consumer's participation in the purchase of products like tea, coffee, and other luxuries, there was no obvious advantage to being able to tell somebody after the fact that you attended a balloon experiment.[13] It is possible, of course, that people held onto their tickets as proof; but barring this evidence, any reader of one of the many accounts of the launch could have claimed attendance. At the same time, many people, because of the nature of ascensions, had not need to buy a ticket and so would no have had any evidence that they attended. Unlike other events for which tickets were required, such as plays, it was extremely difficult to control who got to see the balloon, particularly once it had been launched. In addition, while plays might run over several nights or even over weeks and months, launches typically offered one chance to watch.

Explanations for why people attended launches, then, does not settle well with either a theory of emulation and conspicuous consumption or a notion of consumption as the purchase of signifiers. Another approach one might follow is that of Colin Campbell and his idea of a romantic ethic of consumption. In this instance, science as entertainment appeals to individuals as a sort of '"mentalistic" hedonism' that plays on the romantic (and increasingly Romantic) notion of the balloonist who encouraged public involvement in this new business.[14] The emphasis on the sensations surrounding balloons – the way spectators, and especially women, swooned at the launch, the tears of joy that gushed from the eyes of viewers, the awe-inspiring nature of the event coupled with band music, or perhaps complete silence – suggest that emerging Romanticism played a role here. One spectator at a launch in Lyon, for example, made a point of describing the intense emotions of the audience, which were a jumble of 'joy and fear'.[15] As Campbell argues, historians can understand the motives and intentions which guided eighteenth-century consumers by exploring the written record to which they had access; thus, even if those individuals did not leave us their actual words to analyse, we can at least explore the range of potential meanings and motives which they could have drawn on to construct their meaning from the event.[16] But if that is the case, the romantic ethic appears, when it is apparent at all, almost exclusively in the accounts of balloons written after the launch and not in the advertisements designed to lure prospective customers to purchase tickets. Thus, the romantic ethic as defined by Campbell is most useful to help us understand the creation of ideas about balloon launches after the fact; but it does little to help us understand why people participated in the consumption of balloons to begin with.[17]

This relationship is, of course, more complex. If evidence existed showing people read romantic accounts of aeronautics and only then decided to attend a launch, then there would be a clearer image of the influence of the romantic ethic. If, on the other hand, people became interested in ballooning only after hearing more straightforward accounts, then the Enlightenment influence might hold more sway over people's potential interest in this subject. Since the reality was that both enlightened and romantic accounts circulated simultaneously, it stands to reason that either or both views of ballooning drew in the passions and the interests of the consumers of this new science. As we have seen, some aeronauts touted the utility and science of their inventions while others focused on the variety of entertainments associated with the launch. Still other balloonists concentrated on themselves, the heroic figure of the lone man or woman facing the dangers of the heavens, made all the more real after the first fatalities occurred.

In this way, aeronautics offers a glimpse into the transition from Enlightenment to Romanticism, a shift that in the case of ballooning succeeded mostly in muddling both categories and blurring the lines that historians would like to draw. Indeed, while a transition may have occurred, the evidence from ballooning suggests that, at least in this instance, the two movements were frequently conflated. Thus, the ways in which aeronautics drew on both traditions has enabled historians to approach this topic from either end of the spectrum and find evidence for both the scientific nature of ballooning as well as their emotional side with man engaging in an epic battle against the natural world.

The combination of Enlightenment and Romantic ideas about ballooning indicates a rationale for some of the ways in which individuals and states approached this subject. After their invention, many savants argued strongly and vehemently for the utility of aeronautics. That balloons remained popular in spite of their ultimate failure to achieve any sort of scientific function clearly indicates how much they also appealed to other aspects of the mentality of the times. Similarly, states exerted some control over launches while at the same time appropriating aeronautics into civic festivals. Although dangerous and unable to be steered, the state identified balloons as useful to them because of the emotional response people expressed. The state recognized the value of popular interest in ascensions and tapped into that response with the hope that associating it with a state-sponsored event would add to their popularity among the general public.

In a similar fashion, merchants turned their products into 'aerostatic' versions of their former selves in order to cash in on this trend.

In this manner, balloons are good to think on as they infiltrate so many different aspects of late eighteenth- and early nineteenth-century culture. The science of aeronautics and the wonders of ballooning captured such a large audience in part, perhaps, because of these multiple associations. Replete with potential, balloons at a minimum succeeded as a form of entertainment. Social and cultural, political and economic, balloons had an enormous impact beyond their direct usefulness but one in direct correlation to their ability to astound and amaze..

NOTES

Introduction

1. *Morning Post*, 19 December 1783, n.p. Periodicals cited without page number come from D. Lysons *Collectanea: Or, A Collection of Advertisements and Paragraphs from the Newspapers, Relating to Various Subjects*, 5 vols (Strawberry Hill: T. Kirgate, 1840), a collection of newspaper clippings culled from British periodicals. These always appear in the notes without page numbers.
2. *Gentleman's Magazine*, June 1784, p. 471.
3. *Journal de Paris*, 12 March 1785, pp. 291–2; ibid., 14 March 1785, p. 301.
4. *Journal encyclopédique*, January 1785, p. 293.
5. *Gentleman's Magazine*, December 1811, p. 574.
6. For comparisons to explorers and other famous people see, for example, *Lettre à M. de *** sur son projet de voyager avec la Sphere Aérostatique de M. de Montgolfier* (Paris: Marchands de Feuilles Volonts, n.d), p. 3; Madame la Vicomtesse de Fars Fausse-Landry, *Mémoires de Madame la Vicomtesse de Fars Fausse-Landry, ou Souvenirs d'une octogénaire*, 2 vols (Pairs: Ledoyen, 1830), 179–80; and E. duc de Croy, *Journal inédit du duc de Croy, 1718–1784*, 4 vols, ed. Vicomte de Grouchy and P. Cottin (Paris: Flammarion, 1906), vol. 4, p. 307. For Cook see C. M. Wieland, *Sammtliche Werke*, 42 vols (Leipzig: G. J. Göschen, 1794–1801), vol. 30, p. 3; on Columbus see the anonymous poem quoted in V. Lunardi, *An Account of the First Aerial Voyage in England* (London, 1784), p. 66.
7. S.-N.-H. Linguet (ed.), *Annales politiques, civiles et littéraires du dix-huitième siècle*, 19 vols (London: 1777–92), vol. 10, p. 425.
8. On the comparison with inoculation see *Discours sur les découvertes en general, et particulièrement sur deux des principales découvertes de ce siècle* (Paris: Philippe-Denys Pierres, 1784).
9. On electricity and longitude see *Lettre à M. de *** sur son projet de voyager avec la Sphere Aérostatique de M. de Montgolfier* (Paris: Marchands de Feuilles Volonts, n.d), pp. 3–4.
10. M. Tourneux (ed.), *Correspondance littéraire, philosophique et critique*, 16 vols (Paris: Garnier frères, 1877–82), vol. 13, p. 344.
11. *Journal de Paris*, 30 June 1784, p. 780.
12. T. Cavallo, *The History and Practice of Aerostation* (London, 1784).
13. G. Tissandier, *Histoire des ballons et des aeronauts célèbres* (Paris: Launette, 1887); and M. Mason, *Aeronautica; or, Sketches Illustrative of the Theory and Practice of Aerostation* (London: F.C. Westley, 1838). Also see J.-F. Depuis-Delcourt, *Nouveau manuel complet d'aérostation, ou Guide pour server á l'histoire et á la pratique des ballons* (1850; Paris:

Chez Leonce Laget, 1978); and C. H. Turnor, *Astra Castra: Experiments and Adventures in the Atmosphere* (London: Chapman and Hall, 1865).

14. L. T. C. Rolt, *The Aeronauts: A History of Ballooning, 1783–1903* (New York: Walker, 1966).
15. See, for example, J. Christopher, *Riding the Jetstream: The Story of Ballooning from Montgolfier to Breitling* (London: John Murray, 2001); F. M. Feldhaus, *Luftfahrten einst und jetzt* (Berlin: H. Paetel, 1908); B. Korzus and B. Leismann, eds. *Leichter als Luft: Zur Geschichte der Ballonfahrt* (Münster: Landschaftverband Westfalen-Lippe, 1978); J. Lecornu, *La navigation aérienne: Histoire documentaire et anecdotique*, 3rd edn (Paris: Vuibert & Nony, 1910); and H. Straub, *Fliegen mit Feuer und Gas: Die Geschichte der Ballon- und Luftschiffahrt* (Aarau: AT Verlag, 1984).
16. C. C. Gillispie, *The Montgolfier Brothers and the Invention of Aviation: 1783-1784* (Princeton, NJ: Princeton University Press, 1983). Also see M. Faure, *Les frères Montgolfier et la conquête de l'air* (La Calade, Aix-en-Provence: Edisud, 1983); and P.-L. Clement, *Les Montgolfiers: Leur invention, leur evolution du XVIIIe à nos jours* (Paris: Tardy, 1982).
17. M. Thébaud-Sorger, *L'Aérostation au temps des Lumières* (Rennes: Presses universitaires de Rennes, 2009).
18. T. D. Crouch, *The Eagle Aloft: Two Centuries of the Balloon in America* (Washington, DC: Smithsonian Institution, 1983).
19. J. E. Hodgson, *The History of Aeronautics in Great Britain* (London: Oxford University Press, 1924). On Great Britain also see L. Byrne, *History of Aviation in Ireland* (Tallagh: Folens, 1980).
20. D. Arecco, *Mongolfiere, science et lumi nel tardo settecento* (Bari: Cacucci, 2003). Other works also focus on all or part of the Italian peninsula; see R. Abate, *Storia della aeronautica italiana* (Milan: Bietti, 1974); R. Ambrosini, *L'Aereonautica a Bologna, appunti di cronica raccolti* (Bologna: Tipografia di Paolo Neri, 1912); and G. Boffito, *Il volo in Italia: Storia documentata e aneddotica dell'aeronautica e dell'aviazione in Italia* (Firenze: G. Barbera, 1921).
21. J. Penny, *Up, Up and Away: An Account of Ballooning in and around Bristol and Bath, 1784–1999* (Bristol: Bristol Branch of the Historical Association, 1999). Other regional and national studies include P. Boyé, *Les premières expériences aérostatiques faites en Lorraine (1783–1788)* (1909; Lacour: Nîmes, 1996); F. Digonnet, *L'Invention de l'aérostation à Avignon en 1782 et les premières ascensions dans cette ville* (Avignon: Seguin, 1906); M. Rocamora, *Historia de la navegación aérea en Barcelona* (Barcelona: José Porter, 1948); and P. Vindel, *Las primeras aeronaves en America fueron de invencion Espanola (Mexico, 1784–1785)* (Madrid, 1954).
22. For a comparison between Great Britain and France see R. Gillespie, 'Ballooning in France and Britain, 1783–1786', *Isis*, 75 (1984), pp. 249–68.
23. On the Montgolfiers see Gillispie, *The Montgolfier Brothers* and Faure, *Les frères Montgolfier*, as well as J. Anglade, *Les Montgolfiers* (Paris: Perrin, 1990). On Blanchard see L. Coutil, *Jean-Pierre Blanchard, physicien-aéronaute* (Evreux: Charles Herissey, 1911); on Zambecarri see T. G. Caproni, A. Bertanelli and G. Boffito, *Francesco Zambeccari, aeronauta (Bologna, 1752–1812)* (Milan: Museo Caproni, 1931); on Pilâtre de Rozier see A. Dollfus, *Pilâtre de Rozier: premier navigateur aérien, première victime de l'air* (Paris: Association Française pour l'Avancement des Sciences, 1993); on Tytler see J. Fergusson, *Balloon Tytler* (London: Faber and Faber, 1972); on Lunardi see L. Gardiner, *Man in the Clouds: The Story of Vincenzo Lunardi* (Edinburgh: Chambers, 1963); and G. Morazzoni, *Un pioniere dell'aereonautica, Vincenzo Lunardi* (Milan: W. Toscanini, 1931);

on Garnerin see C. Perrin, *La vie rocambolesque d'André Garnerin, Pionnier du parachute* (Paris: Messene, 2000); and M. Poniatowski, *Garnerin, le premier parachutiste de l'histoire* (Paris: Albin Michel, 1983); on Reichard see H. Monjau, *Wilhelmine Reichard: erste deutsche Ballonfahrerin: 1788 bis 1848: eine dokumentarische Biographie* (Freital: Monjau, 1998).

24. G. Boffito, *Biblioteca aeronautica italiana* (1929; Firenze: L.S. Olschki, 2002); G. Tissandier, *Bibliographie aeronautique* (Paris: Launette, 1887); Vindel, *Historia bibliogáfica*; and A. Wouwermans, *Contribution à la bibliographie de la locomotion aérienne* (Anvers, 1894).
25. L. Weatherill, 'The Meaning of Consumer Behaviour in Late Seventeenth- and Early Eighteenth-Century England', in J. Brewer and R. Porter (eds), *Consumption and the World of Goods*, (London: Routledge, 1993), pp. 206–27. Cf. J. Baudrillard, *The Consumer Society: Myths and Structures* (London: Sage, 1998).
26. Rolt, *The Aeronauts*, pp. 117–18.
27. D. Roche, *A History of Everyday Things: The Birth of Consumption in France, 1600–1800*, trans. B. Pearce (Cambridge: Cambridge University Press, 2000), p. 49.
28. W. Clark, J. Golinski, and S. Schaffer, *The Sciences in Enlightened Europe* (Chicago, IL: University of Chicago Press, 1999), pp. 24–5.
29. On the periodical press see G. Feyel, 'Négoce et presse provinciale en France au 18e siècle: methods et perspectives de recherches', in F. Angiolini and D. Roche (eds), *Cultures et formations négociantes dans l'Europe moderne* (Paris: Editions de l'Ecole des Hautes Etudes en Sciences Sociales, 1995), pp. 439–511; and J. Sgard (ed.), *La presse provinciale au XVIIIe siècle* (Grenoble: Centre de Recherches sur les Sensibilités, 1983).
30. J. de Vries, *The Industrious Revolution: Consumer Behavior and the Household Economy, 1650 to the Present* (Cambridge: Cambridge University Press, 2008), pp. 122–3.

1 The Emergence of Aeronautics

1. Rolt, *The Aeronauts*; Hodgson, *The History of Aeronautics in Great Britain*; Crouch, *The Eagle Aloft*; M. Thébaud-Sorger, *L'Aérostation au temps des Lumières* (Rennes: Presses universitaires de Rennes, 2009); Arecco, *Mongolfiere, science et lumi nel tardo settecento*; Penny, *Up, Up and Away*; B. T. Brown, 'French Scientific Innovation in Late-Eighteenth-Century Dublin: The Hydrogen Balloon Experiments of Richard Crosbie (1783–1785)', in G. Gargett and G. Sheridan (eds), *Ireland and the French Enlightenment, 1700–1800* (New York: St Martin's, 1999), pp. 107–26; P. Holl, 'Les débuts de l'aviation à Strasbourg', *La Vie en Alsace* (1926), pp. 137–42; Boyé, *Les premières expériences aérostatiques faites en Lorraine*; M. Lantier, 'Retombées aérostatiques sur la généralité de Caen (1783–1785)', *Annales de Normandie*, 30 (1980), pp. 271–6.
2. Gillispie, *The Montgolfier Brothers*. Also see J. Anglande, *Les Montgolfiers* (Paris: Perrin, 1990); Clement, *Les Montgolfiers*; and M. Faure, *Les frères Montgolfier et la conquête de l'air* (La Calade, Aix-en-Provence: Edisud, 1983). For other aeronauts see, for example, Dollfus, *Pilâtre de Rozier*; R. Fontaine, *La Manche en ballon: Blanchard contre Pilâtre de Rozier* (Dunkerque : Westhoek-Editions, 1982); Fergusson, *Balloon Tytler*; Poniatowski, *Garnerin*; Perrin, *La vie rocambolesque d'André Garnerin*; and Gardiner, *Man in the Clouds*.
3. W. S. Baring-Gould and C. Baring-Gould, *The Annotated Mother Goose* (New York: New American Library, 1967). I would like to thank Kerry Baxter for providing me with this reference.

4. A. Gellius, *The Attic Nights*, ed. J. C. Rolfe, 3 vols (London: William Heinemann, 1927), vol. 2, p. 245; J. Duhem, *Histoire des idées aéronautiques avant Montgolfier* (Paris: Fernand Sorlot, 1943), p. 123.
5. See Duhem, *Histoire des idées aéronautiques*; Abate, *Storia della aeronautica italiana*, pp. 19–25; and A. Heggen, 'Die 'ars volandi' in der Literatur des 17 und 18 Jahrhunderts', *Technikgeschichte*, 42 (1975), pp. 327–37. An excellent discussion of a number of French authors who discuss flight can be found in A. Coley, 'Followers of Daedalus: Science and Other Influences in the Tales of Flight in Eighteenth-Century French Literature', *Studies on Voltaire and the Eighteenth Century*, 371 (1999), pp. 81–173. For Great Britain see M. H. Nicolson, *Voyages to the Moon* (New York: Macmillan, 1949); and idem., *A World in the Moon: A Study of the Changing Attitude Toward the Moon in the Seventeenth and Eighteenth Centuries* (Northhampton, MA: Smith College, 1936).
6. S. Cyrano de Bergerac, *Voyages to the Moon and the Sun*, trans. R. Aldington (New York: Orion, 1962).
7. François Marie Arouet de Voltaire, 'Micromégas', in *Candide and Other Stories*, trans. Roger Pearson (Oxford: Oxford University Press, 1990), pp. 101–21; J. Galien, *L'art de naviger dans les airs*, 2nd edn (Avignon: Fez, 1757); and Domingo Gonsales, *The Man on the Moone: or A Discourse of a Voyage thither* (London: John Norton, 1638).
8. J.-J. Rousseau, *Le Nouveau Dédale* (Pasedena, CA: The Institute of Aeronautical History, 1950). The anonymous introduction to the modern edition of this work suggests that Rousseau, who wrote this book when he was comparatively young, drew upon the works of seventeenth-century savants Alfonso Borelli and, especially, Honoratus Fabri. See 'Introduction', Rousseau, *Le Nouveau Dédale*, pp. 11–13.
9. N. E. Restif de la Bretonne, *La Découverte australe*, 2 vols (Leipzig, 1781).
10. S. Johnson, *The History of Rasselas, The Prince of Abissinia*, ed. J. P. Hardy (Oxford: Oxford University Press, 1999), ch. 6. On Johnson and his attitudes toward flight see J. E. Hodgson, *Doctor Johnson on Ballooning and Flight* (London: Elkin Mathews, 1925); and L. A. Landa, 'Johnson's Feathered Man: "A Dissertation on the Art of Flying" Considered', in W. H. Bond (ed.), *Eighteenth-Century Studies in Honor of Donald F. Hyde* (New York: The Grolier Club, 1970), pp. 161–78. Cf. S. Aravamudan, *Tropicopolitans: Colonialism and Agency, 1688–1804* (Durham, NC: Duke University Press, 1999), pp. 202–14.
11. S. Johnson, *The Rambler*, 3 vols (London: Thomas Tegg, 1822–1826), vol. 3, p. 304.
12. J. Swift, *Gulliver's Travels* (Indianapolis, IN: Bobbs-Merrill, 1963); A. Behn, 'The Emperor of the Moon', in J. Spencer (ed.), *The Rover; The Feigned Courtesans; The Lucky Chance; The Emperor of the Moon* (Oxford: Clarendon, 1995), pp. 271–335; and D. Defoe, *The Consolidator, or Memoirs of Sundry Transactions from the World in the Moon*, ed. M. J. Bosse (New York: Garland, 1972).
13. Duhem, *Histoire des idées aéronautiques*, pp. 413–17. On early helicopters see J. R. Chiles, *The God Machine: From Boomerangs to Blackhawks: The Story of the Helicopter* (New York: Bantam, 2007), ch. 2.
14. On Burattini see K. Targosz, '"Le dragon Volant" de Tito Livio Burattini', *Annali dell'Istituto e Museo di Storia della Scienza di Firenze*, 2:2 (1977), pp. 67–85; and R. Taton, 'Nouveau document sur le 'dragon volant' de Burattini', *Annali dell'Istituto e Museo di Storia della Scienza di Firenze*, 7:2 (1982), pp. 161–8.
15. See Piero E. Ariotti, 'Christiaan Huygens: Aviation Pioneer Extraordinary', *Annals of Science*, 36 (1979), pp. 611–24.

16. D. Bourgeois, *Recherches sur l'art de voler, depuis la plus haute antiquité jusqu'à ce jour* (Paris: Cuchet, 1784).
17. Launoy and F. Bienvenu, *Instruction sur la nouvelle Machine inventée par MM. Launoy, Naturaliste, & Bienvenu, Machiniste-Physicien* (N.p., n.d.). On Bienvenu see P. Bret, 'Un bateleur de la science: Le "machiniste-physicien" François Bienvenu et la diffusion de Franklin et Lavoisier', *Annales historiques de la Révolution française*, 338:4 (2004), pp. 95–127.
18. On Blanchard see Coutil, *Jean-Pierre Blanchard*.
19. *Journal de Paris*, 28 August 1781, pp. 966–7. See Duhem, *Histoire des idées aéronautiques*, pp. 174–7.
20. *Journal de Paris*, 1 May 1782, p. 483.
21. Ibid., 4 May 1782, p. 495.
22. Ibid., 8 May 1782, p. 511.
23. On Bléton see M. R. Lynn, *Popular Science and Public Opinion in Eighteenth-Century France* (Manchester: Manchester University Press, 2006), ch. 5.
24. Ibid., 22 August 1782, p. 816.
25. Rolt, *The Aeronauts*.
26. [L. G. Gerard], *Essai sur l'art du vol aérien* (Paris: Duchesne, 1784), p. 39; Duhem, *Histoire des idées aéronautiques*, pp. 220–33.
27. Hodgson, *The History of Aeronautics in Great Britain*, pp. 88–92.
28. S. Shapin, *A Social History of Truth: Civility and Science in Seventeenth-Century England* (Chicago, IL: University of Chicago Press, 1995); and L. Hilaire-Pérez, *L'invention technique au siècle des Lumières* (Paris: Albin Michel, 2000).
29. J. V. H. Melton, *The Rise of the Public in Enlightenment Europe* (Cambridge: Cambridge University Press, 2001).
30. Cavallo, *The History and Practice of Aerostation*, pp. 108–79. Cavallo does make some references to Italian launches as well as some brief mention of Switzerland, Spain and the United States. On Dinwiddie see L. Lunney, 'The Celebrated Mr. Dinwiddie: An Eighteenth-Century Scientist in Ireland', *Eighteenth-Century Ireland*, 3 (1988), pp. 69–88.
31. *Journal de Paris*, 11 September 1783, p. 1047; ibid., 12 September 1783, p. 1050. For the repeat performance, ibid., 29 September 1783, p. 1124.
32. F.-L. Ehrmann and C. Würtz, *Montgolfier'sche Luftkörper oder aerostatische Maschinen, eine Abhandlung, worinn die Kunst sie zu verfertigen und die Geschichte der bisher damit angestellten Versuche beschrieben warden* (Strasbourg: Treuttel, 1784); C. Kramp, *Geschichte der aerostatic, historich, physich und mathematisch ausgefuhrt*, 2 vols (Strasbourg, 1784). On ballooning in Strassbourg see Thébaud-Sorger, *L'Aérostation au temps des Lumières*, pp. 158–64.
33. Holl, 'Les débuts de l'aviation à Strasbourg', *La Vie en Alsace* (1926), pp. 139–40.
34. Ibid., p. 140.
35. B. Faujas de Saint-Fond (ed.), *Description des expériences aérostatiques de MM. De Montgolfier*, 2 vols (1784; Osnabrück: Otto Zeller, 1968), vol. 2, pp. 279–81.
36. For Guyton de Morveau's launch see L.-B. Guyton de Morveau, *Description de l'aérostate* (Dijon: Causse, 1784).
37. On ballooning in Saint Domingue see J. E. McClellan III, *Colonialism and Science: Saint Domingue in the Old Regime* (Baltimore, MD: The Johns Hopkins University Press, 1992), pp. 169–70.

38. Faujas de Saint-Fond (ed.), *Description des expériences aérostatiques*, vol. 2, pp. 193–9; Cavallo, *The History and Practice of Aerostation*, pp. 109–10; and Arecco, *Mongolfiere, science et lumi*, pp. 126–7.
39. W. N. Hargreaves-Mawdsley, *The English Della Cruscans and their Time, 1783–1828* (The Hague, Martinus Nijhoff, 1967), pp. 70–1; E. Cochrane, *Tradition and Enlightenment in the Tuscan Academies, 1690–1800* (Chicago, IL: University of Chicago Press, 1961), p. 149. Also see Arecco, *Mongolfiere, science et lumi*, pp. 179–83.
40. Arecco, *Mongolfiere, science et lumi*, pp. 155–6.
41. 'Expérience faite à Milan' in Faujas de Saint-Fond (ed.), *Description des expériences aérostatiques*, vol. 2, p. 158.
42. Cavallo, *The History and Practice of Aerostation*, p. 124.
43. Agostino Gerli, quoted in Anon. (ed.), *The Romance of Ballooning: The Story of the Early Aeronauts* (New York: Viking, 1971), p. 33.
44. Rolt, *The Aeronauts*, p. 57. A. Gerli, *Opuscoli di Agostino Gerli* (Parma: Dalla Samperia Reale, 1785); A. Gerli, G. Gerli and C. G. Gerli, *Maniera di migliorane e dirigere I alloni aerie* (Rome, 1790).
45. *Giornale aerostatico*, 1784.
46. C. Amoretti, *Delle machine aerostatiche*, ed. L. Pescasio (Mantova: Circolo Bibliofilo Aeronautico; Editoriale Padus, 1970). On Amoretti see Arecco, *Mongolfiere, science et lumi*, pp. 157–75.
47. For Zambeccari in Bologna, see his *Relazione dell'esperienza aereo-statica* (Bologna, 1804); and Ambrosini, *L'Aereonautica a Bologna*. On Lunardi in Naples see Arecco, *Mongolfiere, science et lumi*, 190–4; and Gardiner, *Man in the Clouds*, pp. 168–71
48. On early ballooning in Spain see I. Vallès I Rovira, *La Màgia del Vol: Primeres Proves Aerostàtiques a Barcelona, València I Castella al Final des Segle XVIII* (Barcelona: Alta Fulla, 1985); and Tissandier, *Histoire des ballons*, p. 131.
49. Gardiner, *Man in the Clouds*, pp. 172–4.
50. On ballooning in Oaxaca, see V. G. Claveran, 'Globos aerostáticos en la Oaxaca del siglo XVIII', *Quipu*, 4 (1987), pp. 387–400.
51. G. Kuhn, 'Fiestas and Fiascoes – Balloon Flights in Nineteenth-Century Mexico', *Journal of Sport History*, 13:2 (1986), pp. 113–14. On ballooning in Latin America more generally see G. Kuhn, 'The History of Aeronautics in Latin America' (unpublished Ph.D. dissertation University of Minnesota, 1965).
52. J. Alexander, 'Aeromania, "Fire-Balloons," and Catherine the Great's Ban of 1784', *Historian*, 58 (1996), pp. 502–8.
53. Straub, *Fliegen mit Feuer und Gas*, p. 56.
54. Hodgson, *The History of Aeronautics in Great Britain*, pp. 19–20.
55. For Belgium and Holland see P. Verhaegen, 'Les premiers ballons en Belgique, 1783–1786', *Annales de la Société royale d'archéologie de Bruxelles*, 13 (1899): 361–72; and Hodgson, *The History of Aeronautics in Great Britain*, p. 19.
56. Cavallo, *The History and Practice of Aerostation*, p. 135. On Dinwiddie see L. Lunney, "The Celebrated Mr. Dinwiddie: An Eighteenth-Century Scientist in Ireland," *Eighteenth-Century Ireland*, 3 (1988): 69–88.
57. Hodgson, *The History of Aeronautics in Great Britain*, pp. 101–2.
58. Ibid., pp. 103–4.
59. Rolt, *The Aeronauts*, p. 63.
60. On Tytler see Fergusson, *Balloon Tytler*.

61. On Lunardi, and the troubles he faced in attempting to do this launch, see Hodgson, *The History of Aeronautics in Great Britain*, pp. 117–39; and Gardiner, *Man in the Clouds*, pp. 28–43.
62. On Sadler see Hodgson, *The History of Aeronautics in Great Britain*, pp. 140–59; and C. Canby, 'James Sadler, Aeronaut', *British Heritage*, 1:4 (1980), pp. 18–29.
63. Cavallo, *The History and Practice of Aerostation*, p. 176.
64. On Crosbie see Hodgson, *The History of Aeronautics in Great Britain*, pp. 187–9. On Potain see Potain, *Relation aérostatique dédiée à la nation Irlandaise* (Paris: Delaunay, 1824).
65. *Virginia Journal*, 24 June 1784, [p. 2]; *Scots Magazine*, September 1784, p. 496; Cavallo, *The History and Practice of Aerostation*, pp. 151–2. On Carnes see R. D. Hill, 'The Search for Peter Carnes', *Richmond County History*, 10:2 (1978), pp. 5–10; and D. Freeman, 'Riding the Wind in Charliers and Montgolfiers', *Early American Life*, 8:4 (1977), pp. 34–7, 64–5.
66. C. V. Glines, 'First in America's Skies', *Aviation History*, 7 (1996), pp. 26–32; and J.-P. Blanchard, *Journal of My Forty-Fifth Ascension and the First in America*, ed. Carroll Frey (Philadelphia, PA: The Penn Mutual Life Insurance Company, 1943).
67. This would fit it with Max Weber's thesis about the Protestant origins of capitalism. See M. Weber, *The Protestant Ethic and the Spirit of Modern Capitalism*, trans. Talcott Parsons (New York: Scribner's, 1958).
68. Melton, *The Rise of the Public*.
69. There are numerous summaries of this event; this brief description is drawn from Gillispie, *The Montgolfier Brothers*, pp. 118–20. For the last words of Romains see Rolt, *The Aeronauts*, p. 93. Gillispie makes the point that Pilâtre de Rozier probably did not invent the idea of a tandem balloon; Gillispie, *The Montgolfier Brothers*, p. 119.
70. Rolt, *The Aeronauts*, p. 93.
71. *Morning Post*, 17 September 1784, n.p; also *Bath Chronicle*, 30 September 1784, p. 1; and reprinted in *Political Repository*, 14 July 1801, [p. 1].
72. Rolt, *The Aeronauts*, p. 75.
73. *Courier de l'Europe*, 26 September 1786, p. 196.
74. Several accounts mention the presence of the parents. Leslie Gardiner also mentions the presence of the fiancé and claims the young couple was to be married the next day. See Gardiner, *Man in the Clouds*, p. 156.
75. *Scots Magazine*, 48 (1786), pp. 457–8.
76. *Gazetteer and New Daily Advertiser*, 25 June 1785, n.p. For more on this voyage see Potain, *Relation aérostatique*.
77. *New York Packet*, 29 September 1785, n.p.
78. *Morning Herald*, 22 April 1785, n.p.
79. *Scots Magazine*, September 1784, p. 495.
80. *Mail*, 1 October 1792, p. 2.
81. Rolt, *The Aeronauts*, p. 80.
82. Ibid., pp. 94–5.
83. Ibid., p. 95.
84. Ibid., p. 96.
85. Anon., *The Romance of Ballooning*, p. 81.
86. *Providence Gazette*, 26 November 1785, [p. 1]; *Gentleman's Magazine*, September 1785, 744; and Rolt, *The Aeronauts*, pp. 105–6.
87. *Portsmouth Oracle*, 29 August 1818, [p. 3].

88. *Morning Chronicle*, 10 December 1802, [p. 2]; *National Aegis*, 29 December 1802, [p. 4].
89. Cavallo, *The History and Practice of Aerostation*, pp. 111–12.
90. *Morning Post*, 5 October 1784, n.p. Also see Fergusson, *Balloon Tytler*, pp. 80–1.
91. *Boston Patriot*, 11 October 1817, [p. 2].
92. *Gentleman's Magazine*, June 1806, pp. 569–70; *Daily Advertiser*, 2 July 1806, [p. 2].
93. *Courier*, 23 February 1803, [p. 1]; *Evening Post*, 9 February 1803, [p. 2]; Mason, *Aeronautica*, p. 270; and *Gentleman's Magazine*, December 1802, p. 1149.
94. *Gentleman's Magazine*, November 1812, p. 493.
95. *Newburyport Herald*, 2 February 1813, [p. 3]; *National Advocate*, 20 January 1813, [p. 2]; and *New-York Gazette*, 7 January 1813, [p. 2].
96. Rolt, *The Aeronauts*, p. 101.
97. For the false reports of her death see *Gentleman's Magazine*, December 1811, p. 574. On her passing out see Ibid., August 1811, p. 179; *Traveller*, 8 October 1811, [p. 3]; and *Gentleman's Magazine*, March 1812, p. 280.
98. On Marie-Madeleine-Sophie Blanchard see M. Poterlet, *Notice sur Madame Blanchard, aéronaute* (Paris: Imprimerie de Fain, 1819); Coutil, *Jean-Pierre Blanchard*, pp. 20–4; G. Ferrero, 'Sophie Blanchard, Amazzone del Cielo, in Val Trebbia', *Studi genuensi*, 14 (1998), pp. 21–49; and R. R. Schneider, 'Star Balloonist of Europe: The Career of Marie-Madeleine Blanchard', *Consortium on Revolutionary Europe, 1750–1850: Proceedings* (1983), pp. 697–711.
99. *Morning Post*, 18 May 1785, n.p.
100. Duhem, *Histoire des idées aéronautiques*, p. 236.
101. Ibid., pp. 237–40.
102. Ibid., p. 243.
103. Gillispie, *The Montgolfier Brothers*, p. 85.
104. Ibid., p. 86. Also, Duhem, *Histoire des idées aéronautiques*, pp. 249–51. The abbé Bertholon also laid claim, albeit obliquely, to the invention of the parachute. See Gillispie, *The Montgolfier Brothers*, pp. 86–7; and Duhem, *Histoire des idées aéronautiques*, pp. 251–5.
105. T. Martyn, *Hints of Important Uses, To be Derived from Aerostatic Globes. With A Print of an Aerostatic Globe, and its Appendages. Originally Designed in 1783*, 2nd edn (London: by the Author, 1784).
106. *Morning Herald*, 8 November 1784, n.p. On the early history of the parachute see C. A. Prieur de la Côte d'Or, 'Note historique sur l'invention et les premiers essays des parachutes', *Annals de Chimie*, 31 (1799), pp. 269–73; and idem., 'Reclamation relative à l'invention des parachutes', *Annals de Chimie*, 36 (1801), pp. 94–9.
107. See, for example, a short pamphlet, undated but published during the French Revolution, in which Blanchard writes of 'a beautiful parachute experiment that I had invented in 1777.' Blanchard, *L'aéronaute Blanchard aux dames de Valognes* (Valognes: Clamorgam, n.d.), n.p.
108. Hodgson, *The History of Aeronautics in Great Britain*, p. 322.
109. *Morning Post*, 4 June 1785, n.p.
110. Ibid., 6 June 1785, n.p.
111. *Journal de Paris*, 29 August 1785, p. 997.
112. *Gentleman's Magazine*, September 1785, p. 742.
113. Ibid., July 1785, p. 567. For the advertisements announcing the jump see J.-P. Blanchard, *Aerostatic Academy* (n.p., n.d.), n.p.; and *Morning Post*, 26 May 1785, n.p.
114. *World*, 6 November 1788, n.p.

115. Ibid.
116. *Federal Gazette*, 6 June 1793, [p. 3]. For announcements concerning the experiment see the *National Gazette*, 20 March 1793, p. 163; and the *General Advertiser*, 14 June 1793, [p. 3].
117. *City Gazette*, 29 June 1793, [p. 2].
118. *Newport Mercury*, 16 May 1796, [p. 3].
119. *Oracle of the Day*, 22 March 1797, [p. 3]; *Minerva*, 6 March 1797, [p. 3].
120. *Gentleman's Magazine*, June 1785, p. 484; this story is reprinted in the *Diary*, 29 January 1793, [p. 2].
121. *Connecticut Journal*, 15 February 1786, [p. 3].
122. *Carolina Federal Republican*, 23 August 1817, [p. 3].
123. On Garnerin see Perrin, *La vie rocambolesque d'André Garnerin*; and Poniatowski, *Garnerin*.
124. See the autobiographical sketch A.-J. Garnerin, *Voyage et captivité de [Garnerin], ancien commissaire de la République française, prisonnier d'état en Autriche* (Paris: Librairie du Cercle-Social, 1797).
125. Hodgson, *The History of Aeronautics in Great Britain*, pp. 218–19. On Garnerin's loss of his position as imperial aeronaut, see chapter 5.
126. *Chronique de Paris*, 24 June 1790, p. 700.
127. *Journal de Paris*, 11 July 1790, n.p.
128. *Chronique de Paris*, 7 June 1790, p. 630.
129. *Reimpression de l'Ancien Moniteur*, 29 (24 October 1797), p. 46. See the description of this event in Rolt, *The Aeronauts*, p. 106. An alleyway circling the inside fence of Parc Monceau is still named after Garnerin in honour of the occasion.
130. A.-J. Garnerin, *Parc de Mousseaux. Voyage aérien du Citoyen Garnerin, avec une jeune personne* (Paris: Imprimerie du Cercle-Social, An VI).
131. A.-J. Garnerin, *Rapport fait par le C. Garnerin, de son voyage aérien* ([Paris]: Imprimerie-Librairie du Cercle Social, An VI), pp. 1–2.
132. Ibid., p. 4.
133. A.-J. Garnerin, *Physique: Voyage aérien* (N.p., n.d.).
134. In addition to those discussed below see the description of a flight on 28 June 1802 in *Gentleman's Magazine*, July 1802, pp. 663–8.
135. *Gentleman's Magazine*, August 1802, pp. 775–6.
136. Ibid., September 1802, p. 873.
137. A.-J. Garnerin, *Détails des trois premiers voyages aériens que M. Garnerin a fait en Russie* (Moscow: Typographie de l'Université, 1803), pp. 7–9, 33–4.
138. Monck Mason also lists flights by Madame Jacques Garnerin, using the name Labrosse, Mlle. Eugene Garnerin, Mlle. Cécile-Benoit Garnerin, and Mlle Blanche Garnerin, all taking place between 1798 and 1818. See Mason, *Aeronautica*, p. 256.
139. *Descente en parachute par Mlle. Garnerin, Le 25 Octobre 1815* (n.p., n.d.)
140. *Evening Post*, 6 May 1816, [p. 2].
141. *Yankee*, 17 October 1817, [p. 3] (Rouen); J.A. Fortuna, *Exposé des faits relatifs à l'expérience de Mlle. Elisa Garnerin à Madrid* (Paris: A. Bobée, 1818); L. Pecchio, *A Elise Garnerin, au sujet de son vol aérostatique exécute à Milan le avril 1824* (Milan: Nicolas Bettoni, 1824); *Elisa Garnerin nel suo XXIII volo da Milano il XIX Aprile 1824* (Milan: Bettoni, 1824); and *Estatta descrizione del globo e paracadute dell-aeronauta Madamigella Elisa Garnerin in occaione del suo volo eseguito in Padova nell-anno 1825* (Padova: Gamba, n.d.).

142. P. Caron, and C. Gével, 'Mademoiselle Elisa Garnerin, aéronaute', *Revue politique et littéraire: Revue bleue* 14 (8 Avril 1912), pp. 438–9.
143. *Enlevement à ballon perdu d'un physicien qui redescendra en para-chute* (n.p., n.d.); on the Ruggieri brothers see M. R. Lynn, 'Sparks for Sale: The Culture and Commerce of Fireworks in Early Modern France', *Eighteenth-Century Life*, 30 (2006), pp. 74–97.
144. *Connecticut Journal*, 9 January 1800, [p. 1].
145. *New-York Daily Advertiser*, 12 August 1818, [p. 2].

2 The Enlightenment and the Ulitity of Ballooning

1. Lines excerpted from the poem 'On Air Balloons', *Independent Ledger*, 11 October 1784, [p. 4].
2. For the 'sublime invention' see Nicolas Edme Restif de la Bretonne in L. S. Mercier and Restif de la Bretonne, *Paris le jour, Paris la nuit: 'Tableau de Paris', 'Le nouveau Paris' de Louis Sébastien Mercier & 'Les nuits de Paris' de Restif de la Bretonne,* ed. M. Delon and D. Baruch (Paris: R. Laffont, 1990), p. 1051; and for the 'philosophical phenomenon' see *Morning Chronicle*, 26 December 1783.
3. *Lettre à M. de *** sur son projet de voyager avec la Sphere Aérostatique de M. de Montgolfier* (Paris: Marchands de Feuilles Volonts, n.d.), p. 7.
4. J. Sgard, 'Les philosophes en montgolfière', *Studies on Voltaire and the Eighteenth Century*, 303 (1992), p. 99.
5. Walpole cited in Christopher, *Riding the Jetstream*, p. 1.
6. For the 'imperfect state' see *Bath Chronicle*, 14 October 1784, p. 4. J. Jeffries, *A Narrative of Two Aerial Voyages of Dr. Jeffries with Mons. Blanchard; with Meteorological Observations and Remarks* (1786; New York, 1941), pp. 9–10.
7. *Lettre à Mr. M. de Saint-Just, sur le globe aërostatique de MM. Montgolfier* (Paris: Mérigot, Royez, 1784), p. 3. One example of someone who posed the question regarding the use of balloons is X. de Maistre, *Prospectus de l'expérience aérostatique de Chambéry, publié au nom des premiers souscripteurs* (Chambéry: F. Puthod, 1784), p. 5.
8. For Franklin's quip see, among many other sources, Tourneux (ed.), *Correspondance littéraire, philosophique et critique*, vol. 13, p. 349.
9. *Pennsylvania Packet*, 9 December 1783, n.p.
10. *The London Chronicle*, 22–5 January 1785, n.p.
11. D. Fabry, *Réflexions sur la relation du voyage aérien de MM. Charles & Robert, et la brochure intitulée: Méthode aisée de faire la Machine aérostatique* (Paris: Chez les Libraries des Nouveautés, 1784), p. 4.
12. For Britain see the *Morning Herald*, 1 October 1783; for France, *Correspondance littéraire*, 13:344. For all the 'savants', see J.-P. Blanchard, *Procès-verbaux et details des deux voyages aériens faites d'après la découverte de MM. Montgolfier* (Bruxelles: Bailly, 1783), p. 3; for physicists, mechanics, and savants see *Courier de l'Europe*, 9 January 1784, p. 18.
13. On the importance of utility for the Enlightenment see R. Chartier, *The Cultural Origins of the French Revolution*, trans. Lydia G. Cochrane (Durham: Duke University Press, 1991), p. 17; with respect to utility as it relates to utilitarianism and

politics see P. Gay, *The Enlightenment: An Interpretation*, 2 vols (New York: Norton, 1966–9), vol. 2, pp. 416, 458–61; and A. Cobban, *In Search of Humanity: The Role of the Enlightenment in Modern History* (New York: George Braziller, 1960), pp. 126–32.

14. L. Roberts, 'Going Dutch: Situating Science in the Dutch Enlightenment', in W. Clark, J. Golinski and S. Schaffer (eds), *The Sciences in Enlightened Europe*, (Chicago, IL: University of Chicago Press, 1999), pp. 350–88, p. 351.
15. T. Baldwin, *Airopaidia* (Chester: Printed for the Author, 1786), p. 259.
16. *Affiches de Province*, 18 November 1784, p. 650.
17. B. Faujas de Saint-Font, *A Journey Through England and Scotland to the Hebrides in 1784*, ed. A. Geikie, 2 vols (Glasgow: Hugh Hopkins, 1907), vol. 1, p. 25.
18. *Gentleman's Magazine*, June 1785, p. 481.
19. Ibid., December 1788, pp. 1040–1.
20. 'Reflections on Balloons', *National Gazette*, 19 January 1793, 2:24, p. 94.
21. J. Sadler, *Balloon: An Authentic Narrative of the Aerial Voyage of Mr. Sadler, Across the Irish Channel* (Dublin, 1812), pp. 22–3.
22. *Bath Chronicle*, 6 May 1784, p. 4.
23. *The Air Balloon: Or a Treatise on the Aerostatic Globe*, 4 edn (London: Kearsley, 1784), p. 4.
24. Cowper to John Newton, 17 November 1783, in *The Letters and Prose Writings of William Cowper*, ed. J. King and C. Ryscamp, 3 vols (Oxford: Oxford University Press, 1981), vol. 2, pp. 181–2.
25. [Piroux], *L'art de voyager dans les airs, ou les ballons* (Paris, 1784), pp. 41–2. This list of sciences appeared frequently. See also *Journal de Paris*, 17 November 1783, pp. 1320–1; and *Affiches de Province*, 20 December 1783, p. 210.
26. See the Fonds Montgolfier at the Musée de l'Air et l'Espace in Le Borget, France.
27. *Morning Herald*, 11 September 1783. Also see the comments by the anonymous author of *Lettre à Mr. M. de Saint-Just, sur le globe aërostatique de MM. Montgolfier* (Paris: Mérigot, Royez, 1784), p. 9.
28. An unnamed Russian minister cited in N. Galitzyne, 'Les premières expériences de Montgolfier: D'après des documents russes', in *Annales internationals d'histoire. Congrès de Paris, 1900*, 7 vols in 2 tomes (Nendeln/Liechtenstein: Kraus Reprint, 1972), tome II, vol. 5, pp. 146–54, citation from pp. 150–1.
29. Jefferson to Philip Turpin, 28 April 1784, in *Papers*, ed. J. P. Boyd, 10 vols (Princeton, NJ: Princeton University Press, 1950–95), vol. 6, p. 136.
30. P. M. Freneau, *The Poems of Philip Freneau*, ed. Fred Lewis Pattee (Princeton, NJ: The University Library, 1902–07), pp. 276–8. Originally printed in the *Freeman's Journal*, 22 December 1784, [p. 2].
31. M. A. Costa, *Saggi sull'aerostatica e sull'aeronautica* (Naples: Fibreno, 1837).
32. *Lettre à un ami, sur l'utilité des globes volans, de M. de Montgolfier, & sur la possibilité de la prise de Gibralter* (Paris: Gueffier, 1783), pp. 15–18 (astronomy), and pp. 3–10 (Gibralter).
33. Martyn, *Hints of Important Uses* p. 6. Francis Olivari described a similar plan for Ireland in which thirty equidistant balloons would be stationed along the coast with flags to act as telegraph operators and relay important news around the country. See F. Olivari, *An Essay on Aérostation* (Dublin: George Bonham, 1797).
34. Martyn, *Hints of Important Uses*, p. 11.
35. [Gerard], *Essai sur l'art du vol aérien*, pp. 141–2.

36. Ibid., pp. 143–4.
37. Ibid., pp. 145–7.
38. Cavallo, *History and Practice of Aerostation*, p. 190. Emphasis in the original.
39. Ibid., p. 191.
40. Ibid., p. 320.
41. Ibid., pp. 320–2.
42. Ibid., p. 323.
43. J.-P. Blanchard and G. Baker, *The Principles, History and Use of Air-Balloons* (New York: Van Alen, 1796), pp. 25–6. This section is reprinted in the anonymous book *Aerostation Displayed* (London: Hammond, 1802), pp. 50–1.
44. *Morning Post*, 20 September 1784.
45. New Hampshire Gazette, 17 June 1785, [p. 3].
46. *Independent Gazetteer*, 26 June 1784, [p. 2].
47. *Register of the Times*, 23 December 1796, [p. 4].
48. *Hive*, 7 February 1804, p. 47
49. *Daily Advertiser*, 26 October 1804, [p. 2].
50. *Claypoole's American Daily Advertiser*, 2 November 1799, [p. 3].
51. *Poulson's American Daily Advertiser*, 13 November 1805, [p. 2].
52. Gay-Lussac, 'Relation d'un voyage aérostatique', *Journal de physique, de chimie, et d'histoire naturelle* 59 (1804): 454–462; C. Gillispie, *Science and Polity in France: The Revolutionary and Napoleonic Years* (Princeton, NJ: Princeton University Press, 2004), p. 639.
53. R. Schofield, *The Lunar Society of Birmingham: A Social History of Provincial Science and Industry in Eighteenth-Century England* (Oxford: Oxford University Press, 1963), p. 252.
54. Cavallo, *The History and Practice of Aerostation*, pp. 108–9.
55. Jeffries, *A Narrative of Two Aerial Voyages*, pp. 10–11.
56. P. Bertholon, *Des avantages que la physique, et les arts qui en dependent, peuvent retirer des globes aërostatiques* (Montpellier: Jean Martel, 1784), pp. 16–19.
57. J.-P. Blanchard, *Journal and Certificates on the Fourth Voyage of Mr. Blanchard* (London: Baker and Galabin, 1784), p. 2. A flageolet is a musical instrument.
58. Jeffries, *A Narrative of Two Aerial Voyages*, pp. 13–14.
59. *Morning Post*, 7 May 1785.
60. *New York Packet*, 23 February 1786, [p. 2].
61. Rolt, *The Aeronauts*, p. 70.
62. Vincenzo Lunardi to King George III, 16 September 1784, in F. and A. Kermode (eds), *The Oxford Book of Letters*, (Oxford: Oxford University Press, 1995), p. 158.
63. *Gentleman's Magazine*, 58 (1785), pp. 652–3.
64. Blanchard, *Journal of My Forty-Fifth Ascension*, pp. 21–2.
65. Cavallo, *The History and Practice of Aerostation*, pp. 163–4.
66. *Gentleman's Magazine*, 55 (1785), p. 400.
67. H. Beaufoy, *Journal kept by H.B.H.B. during an aerial voyage with Mr. James Sadler* (London: G. Woodfall, 1811), p. 1.
68. *Daily Advertiser*, 4 August 1802, [p. 3].
69. Walpole, quoted in Rolt, *Aeronauts*, p. 159.
70. On Montgolfier and Franklin see Rolt, *Aeronauts*, p. 159. On military ballooning see, among others, J. Godechot, 'L'Aérostation militaire sous le directoire', *Annales historiques de la Révolution française*, 8 (1931), pp. 213–28.
71. *Affiches de Province*, 26 October 1783, p. 1231.

72. *Courrier d'Avignon*, 7 November 1783, p. 356.
73. J. Money, *A Short Treatise on the Use of Balloons and Field Observations in Military Operations* (London: C. Roworth, 1803), pp. 5, 12–13.
74. Jan Ingenhousz, cited in M. E. Connaughton, '"Ballomania", the American Philosophical Society and Eighteenth-Century Science', *Journal of American Culture* 7 (1984), p. 71.
75. *Connecticut Gazette*, 16 April 1800, [p. 3].
76. Cited in Alexander, 'Aeromania', p. 514.
77. *Columbian Centinel*, 4 July 1792, p. 130.
78. *Morning Herald and Daily Advertiser*, 2 September 1784.
79. For dispatches see *Mercantile Advertiser*, 11 April 1799, [p. 3]; for letters and provisions see *State Gazette of South-Carolina*, 30 May 1785, [p. 2].
80. *Reimpression de L'Ancien Moniteur*, 32 vols (Paris: Plon, 1858–63), vol. 24, p. 73 (27 November 1797).
81. *Philadelphia Gazette*, 11 January 1799, [p. 3].
82. Banks to Franklin, 13 September 1783, in *The Letters of Sir Joseph Banks: A Selection, 1768–1820*, ed. N. Chambers (London: Imperial College Press, 2000), p. 62.
83. Linguet (ed.), *Annales politiques*, vol. 11, pp. 34–5.
84. Dorset to Carmarthen, 9 February 1786, in O. Browning (ed.), *Despatches from Paris, 1784–1790*, 2 vols (London: Offices of the Society, 1909–10), vol. 1, pp. 100–1.
85. For the Montgolfier brothers see the Fonds Montgolfier at the Musée de l'Air et d'Espace, Le Borget, France. Also, see Gillispie, *The Montgolfier Brothers and the Invention of Aviation*, pp. 120–9.
86. *Affiches de Province*, 13 January 1784, p. 23.
87. *Essex Journal* [Massachusetts], 12 January 1785, [p. 3].
88. *Journal gratuit*, 6 (1790), pp. 113–15; and *Aréostation. Establissement d'une compagnie aéronautique* (n.p., 1790).
89. *Gentleman's Magazine*, September 1817, p. 252.
90. *Commercial Advertiser*, 8 February 1800, [p. 3].
91. *Public Advertiser*, 1 October 1811, [p. 2].
92. *New York Packet*, 9 May 1785, [p. 2]. As the author describes Vander Monde as an academician, he is perhaps referring to Alexandre-Théophile Vandermonde (1735–96).
93. *New-Haven Gazette*, 24 June 1784, [p. 4].
94. *New-England Gazette*, 3 April 1818, [p. 2].
95. Cowper to Joseph Hill, 22 January 1785, *The Letters and Prose Writings of William Cowper*, vol. 2, p. 321.
96. *Massachusetts Centinel*, 21 April 1784, [p. 4].
97. *Pennsylvania Packet*, 3 April 1784, [p. 2].
98. Ibid., 30 August 1785, [p. 2].
99. *Poulson's American Daily Advertiser*, 24 May 1815, p. 2.
100. *New-Hampshire Gazette*, 17 September 1799, [p. 2].
101. *Connecticut Journal*, 3 June 1801, [p. 3]. Also see *The Bee*, 15 April 1801, [p. 4].
102. *Gazette of the United States*, 9 January 1801, [p. 2].
103. *Massachusetts Centinel*, 23 December 1786, p. 110.
104. *Pennsylvania Packet*, 6 March 1784, [p. 2].
105. *The Morning Post*, 3 December 1784, [p. 2].
106. *Affiches de Toulouse et du Haut Languedoc*, 28 January 1784, pp. 16–17.
107. *Morning Herald and Daily Advertiser*, 20 October 1784, [p. 2]; and ibid., 12 November 1784, [p. 2].

108. Ibid., 13 October 1784, [p. 2]. Also see the poem in *The Gazetteer and New Daily Advertiser*, 3 November 1784, [p. 2].
109. *Morning Herald and Daily Advertiser*, 6 October 1784, [p. 2].
110. *Morning Post*, 15 October 1784, [p. 2].
111. Fabry, *Réflexions sur la relation du voyage aérien de MM. Charles & Robert, et la brochure intitulée*, p. 4.
112. Marat, *Lettres de l'observateur bon-sens, à M. de ***, sur la fatale catastrophe des infortunes Pilatre de Rozier & Romain* (Paris: Méquignon, 1785), p. 32.
113. J. Brown, *Select Remains of the Rev. John Brown* (Pittsburgh, PA: Cramer, Spear and Eichbaum, 1810), p. 107.
114. R. S. Medley, cited in Hodgson, *The History of Aeronautics in Great Britain*, p. 211.
115. John Newton to the Earl of Dartmouth, 22 September 1784, in F. and A. Kermode (eds), *The Oxford Book of Letters*, (Oxford: Oxford University Press, 1995), p. 161.
116. *New-Jersey Gazette*, 16 March 1784, [p. 2].
117. *Pennsylvania Packet*, 22 June 1784, [p. 2].
118. Originally published in the *London Packet*; reprinted in the *Boston Gazette*, 26 April 1784, [p. 4].
119. *Virginia Journal*, 15 July 1784, [p. 2].
120. *Morning Herald and Daily Advertiser*, 8 October 1784, [p. 1].
121. Anonymous letter to M. de Saint-Just, 1784, reprinted in Anon. (ed.), *The Romance of Ballooning*, p. 53.
122. *Columbian Herald*, 23 November 1784, [p. 3].
123. These images can be found in Anon. (ed.), *The Romance of Ballooning*, pp. 56–7.
124. T. Carlyle, *The French Revolution: A History*, rev. edn, 2 vols (New York: Collier, 1900), vol. 1, p. 46.
125. Anna Seward to Humphry Repton, 14 October 1788, in *Letters of Anna Seward*, ed. A. Constable, 6 vols (Edinburgh: Archibald Constable, 1811), vol. 2, p. 174.
126. *South-Carolina Weekly Gazette*, 22 January 1785, [p. 2].
127. D. Bourgeois, *Recherches sur l'art de voler, depuis la plus haute antiquité jusqu'à ce jour* (Paris: Cuchet, 1784), p. 85. This sentiment is repeated by Mathurin-Jacques Brisson in his *Observations sur les nouvelles découvertes aërostatiques et sur la probabilité de pouvoir diriger les Ballons* (Paris: Boucher & Lamy, 1784), p. 13.
128. A.-J. Robert and N.-L. Robert, *Mémoire sur les Expériences Aérostatiques* (Paris: L'Imprimerie de P.-D. Pierres, 1784), p. 3.
129. N. Low, *Low's Almanack and Astronomical and Agricultural Register* (Boston, MA: Munroe & Francis, 1819), n.p.
130. *Morning Chronicle*, 3 September 1803, [p. 2].
131. Thomas Jefferson to Francis Hopkinson, 25 September 1785, in *The Writings of Thomas Jefferson*, ed. A. A. Lipscomb and A. E. Bergh, 20 vols (Washington, DC: Thomas Jefferson Memorial Association, 1903–4), vol. 5, pp. 147–8.
132. On the claims made by Alban and Vallet see the brief summary in *Journal Encyclopédique*, October 1785, p. 135
133. .Ibid., March 1784, pp. 507–8 (Dijols la Castagne); *Morning Post*, 13 December 1784, [p. 1] (Dufour); and *Journal de Paris*, 24 January 1786, p. 98 (Digues).
134. This experiment was reported in *Walker's Hibernian Magazine*, April 1785, p. 191; and *Scots Magazine*, February 1785, p. 75.
135. Blanchard, *Journal of My Forty-Fifth Ascension*, p. 7.
136. *Gentleman's Magazine*, May 1785, pp. 398–99; *Litchfield Monitor*, 26 July 1785, [p. 2].

137. *Morning Post*, 11 May 1785, [p. 1].
138. Etienne Montgolfier's success in getting money from the government was widely reported. See, for example, *Pennsylvania Packet*, 7 May 1785, [p. 2]; more generally, see Gillispie, *The Montgolfier Brothers*, pp. 120–9; and Gillispie, *Science and Polity in France at the End of the Old Regime*, pp. 455–9.
139. On Lunardi see, for example, Vincenzo Lunardi to King George III, 16 September 1784, in *The Oxford Book of Letters*, pp. 158–60.
140. *Journal de Nancy*, 12:18 (1783), pp. 99–100.
141. M. Salle, *Moyen de diriger l'aérostat* (Paris: Couturier, 1784).
142. See the frontispiece of G. D. Bruno, *L'Utile direzione delle machine aerostatiche* (Naples, 1784).
143. J.-L. Carra, *Essai sur la nautique aérienne* (Paris: Eugène Onfroy, 1784).
144. B. Scott, *Aérostat dirigeable à volonté* (Paris: Maradan, 1789).
145. C.G. Kratzenstein, *L'art de naviguer dans l'air* (Copenhave & Leipzig: Faber & Nitschke, 1784), pp. 33–8.
146. C. F. Meerwein, *L'art de voler à la manière des oiseaux* (Basle: J.J. Thourneysen, 1784).
147. Summary of Calvi, *Metodo di dirigere Palloni* (Milan, 1784) in *Monthly Review*, Appendix (1784), p. 584.
148. F. Henin, *Mémoire sur la direction des aerostats* (Paris: Moreau, An X).
149. *New-Haven Gazette and Connecticut Magazine*, 1 June 1786, p. 128.
150. Bourgeois, *Recherches sur l'art de voler*, pp. 85–98.
151. Ibid., p. 86.
152. [Luzarche], *Nouveaux appareils pour la direction des aérostats, ou essai sur cette direction* (Paris: H.-l. Perronneau, 1812).
153. *New-Hampshire Patriot*, 3 September 1816, [p. 3].
154. Renaux to Monsieur le Directeur du Musée à Bordeaux, cited in M. Meaudre de Lapouyade, *Les premiers aeronauts bordelais, 1783–1799* (Bordeaux: Counouilhou, 1910), pp. 51–2.
155. See S. Hoole, *Thoughts on the Farther Improvement of Aerostation* (London: Nicol, 1785), p. 12 (removal of gondola), 15–17 (on shapes), 32 (for the Flying Fish).
156. *Albany Gazette*, 8 August 1815, [p. 3]; *Rhode-Island American*, 9 October 1816, [p. 3]. The engineer named Pauly may be the same person who, along with a certain Frenchman Le Mercier, reported in 1806 their ability to steer balloons. See *Newburyport Herald*, 22 April 1806, [p. 2]. However, another person with a similar name, the Swiss savant named Pauli, also wrote on this topic; see *United States Gazette*, 29 October 1804, [p. 3].
157. *Gentleman's Magazine*, April 1810, p. 372; and *Columbian*, 27 March 1818, [p. 2]. For Degen see J. Degen, *Beschreibung einer neuen flugmaschine* (Vienna, 1808).

3 Balloonists and their Audience

1. *Journal de Paris*, 8 October 1784, pp. 1189–91.
2. *Gentleman's Magazine*, 56 (November 1784), pp. 873–4.
3. *Scot's Magazine*, November 1784, pp. 560–1; *Monthly Review*, October 1784, pp. 383–384.
4. In counting the occupants of the launches I have included everybody named even though it is unclear how much they participated in the launches. Some were probably relatively passive passengers just along for the ride while others may have helped out. This is the case, for example, of the Marquis d'Arlandes who assisted Pilâtre de Rozier. The multiple

members of the 10 January 1784 launch in Lyons, however, probably did little to help plan or execute the launch. I will differentiate between aeronauts and passengers below.

5. Mason, *Aeronautica*; Depuis-Delcourt, *Nouveau manuel complet d'aérostation*, pp. 222–9; Turnor, *Astra Castra*, pp. 457–62. Turnor based his list on Depuis-Delcourt. A more selective list can be found in Korzus and Leismann (eds), *Leichter als Luft*, pp. 284–9.
6. For the story of Wilcox, see Mason, *Aeronautica*, pp. 281–2. On the hoax see Rolt, *The Aeronauts*, p. 140.
7. Turnor, *Astra Castra*, p. 457.
8. Ibid., p. 457.
9. On Marie-Madeleine Sophie Blanchard see Ferrero, 'Sophie Blanchard, Amazzone del Cielo, in Val Trebbia'; and Schneider, 'Star Balloonist of Europe', pp. 696–711.
10. On Wilhelmine Reichard see Monjau, *Wilhelmine Reichard*.
11. *New York Commercial Advertiser*, 22 April 1812, [p. 3].
12. For Aix-la-Chapelle see *Republican Agriculturalist*, 24 December 1818, [p. 1]; for the Hamburg launch see *Boston Intelligencer*, 9 November 1816, [p. 1].
13. Caron, and Gével, 'Mademoiselle Elisa Garnerin.
14. *Estatta descrizione del globo e paracadute dell-aeronauta Madamigella Elisa Garnerin in occaione del suo volo eseguito in Padova nell-anno 1825* (Padova: Gamba, n.d.); and J.A. Fortuna, *Exposé des faits relatifs à l'expérience de Mlle. Elisa Garnerin à Madrid* (Paris: A. Bobée, 1818).
15. See, for example, Louis Pecchio, *A Elise Garnerin, au sujet de son vol aérostatique exécute à Milan le avril 1824* (Milan: Nicolas Bettoni, 1824); and *Elisa Garnerin nel suo XXIII volo da Milano il XIX Aprile 1824* (Milan: Bettoni, 1824).
16. *Descente en parachute par Mlle. Garnerin, Le 25 Octobre 1815* (N.p., n.d.).
17. *Yankee*, 17 October 1817, [p. 3].
18. On Tytler see Fergusson, *Balloon Tytler*.
19. *Morning Post*, 7 January 1785, n.p.; emphasis in original.
20. Ibid., 11 January 1785, n.p.
21. On James Sadler see Canby, 'James Sadler, Aeronaut'.
22. S. Oettermann, 'Die Fliegenden Plastiken des Johann Karl Enslen: Johann Karl Enslen's Flying Sculptures', *Daidalos: Berlin Architectural Journal* 37 (15 September 1990): 44–53.
23. Turnor, *Astra Castra*, p. 457.
24. I borrow this notion from M. Beer, 'Flying High and Living Large: Eighteenth-Century Balloonists and Twenty-First Century Pop Stars.' Unpublished paper presented at the Spring Annual Research Conference, Agnes Scott College, Decatur, GA, April 2004.
25. *Merrimack Magazine*, 4 January 1806, p. 84.
26. Barthélemy Faujas de Saint Fond (ed.), *Description des expériences aérostatiques de MM. De Montgolfier*, 2 vols (Orig. ed. 1784; Osnabrück: Otto Zeller, 1968), vol. 2, p. 42.
27. J.-P. Blanchard, *Relation du quinzieme voyage aërien de M. Blanchard, fait à Francfort sur le Meyn, le 3 Octobre 1785* (Francfort sur le Meyn, 1785), pp. 6–8.
28. V. Lunardi, *An Account of Five Aerial Voyages in Scotland* (London, 1786), p. 105.
29. Ibid., pp. 107–12.
30. Manoel Maria de Barbosa du Bocage, *Elogio Poetico à admiravel intrepdiz, com que en Domingo 24 de Agosto de 1784* (Lisboa: Ferreira, 1794), p. 7.
31. *Morning Post*, 21 September 1784, n.p.
32. Ibid., 1 October 1784, n.p.
33. *Commercial Advertiser*, 16 September 1817, [p. 2].
34. Jeffries, *A Narrative of Two Aerial Voyages*, pp. 71–3.

35. S. Schama, *Citizens: A Chronicle of the French Revolution* (New York: Vintage, 1989), pp. 130–1.
36. M.-E.-G. Duchosal, *Blanchard, poème en quatre chants* (Bruxelles, 1786), p. 63. This is a longer version of a poem that first appeared in 1784; Duchosal, *Blanchard, poème en deux chants* (Rouen; Paris: Marchands de Nouveautés, 1784).
37. *Time Piece*, 15 May 1797, p. 112.
38. M. Luxero, *Portrait de la beauté, et enlèvement de Vénus dans le Vaisseau-volant de M. Blanchard* (Rome; Paris: Chez les Marchands de Nouveautés, 1783). This poem was based on Blanchard's earlier attempt to ascend in a flying boat, an effort that also resulted in other poems such as Verninac de Saint-Maur's *Epitre à Monsieur Blanchard* (Bruxelles, 1782).
39. *Journal de Paris*, 19 October 1784, p 1233.
40. *Chanson du ballon de M. de Montgolfier*. Lyon, [*c*. 1783].
41. Broadside found in the Gimbel Collection, USAF Academy, XC-10–3B (Sad) – 3D.
42. *Précis historique de la grande expérience faite à Lyon la 19 Janvier 1784* ([1784]); *Supplément à l'art de voyager dans les airs, contenant le Précis historique de la grande Expérience faite à Lyon le 19 Janvier 1784 et l'Exposé d'un moyen ingénieux pour diriger a volonté les Ballons aérostatiques* (N.p., n.d.); and R. de Cazenove, *Premiers voyages aériens à Lyon en 1784* (Lyon: Pitrar, 1887).
43. Cavallo, *The History and Practice of Aerostation*, pp. 114–15.
44. Ibid., pp. 122–3.
45. For the inclusion of the animals in the flight see *Newport Mercury*, 6 November 1784, [p. 2]; for their presence at the Pantheon see *Bath Chronicle*, 7 October 1784, p. 2.
46. *Pennsylvania Packet*, 20 November 1784.
47. *Massachusetts Centinel*, 24 November 1784, [p. 1].
48. Blanchard, *Journal of My Forty-Fifth Ascension*, p. 17. The dog also makes an appearance in the children's book by Alexandra Wallner, *The First Air Voyage in the United States: The Story of Jean-Pierre Blanchard* (New York: Holiday House, 1996).
49. *Claypoole's American Daily Advertiser*, 5 July 1797, [p. 3].
50. Jeffries, *A Narrative of Two Aerial Voyages*, p. 21.
51. *Freeman's Journal and Daily Advertiser*, 23–5 June 1785, n.p.
52. F. Venturi, *The End of the Old Regime in Europe, 1776–1789*, trans. R. Burr Litchfield, 2 vols (Princeton, NJ: Princeton University Press, 1991), vol. 2, p. 935.
53. *Morning Post*, 3 November 1784, n.p.
54. On the audience for science and the nature of the audience's relationship to science see Shapin, 'The Audience for Sciences in Eighteenth-Century Edinburgh'.
55. On theatre and opera see J. Ravel, *The Contested Parterre: Public Theater and French Political Culture, 1680–1791* (Ithaca, NY: Cornell University Press, 1999); J. H. Johnson, *Listening in Paris: A Cultural History* (Berkeley, CA: University of California Press, 1996). On fireworks see Lynn, 'Sparks for Sale: The Culture and Commerce of Fireworks in Early Modern France'. Also see S. Werrett, 'From the Grand Whim to the Gasworks: "Philosophical Fireworks" in Georgian England', in L. Roberts, S. Schaffer and P. Dear (eds), *The Mindful Hand: Inquiry and Invention From the Late Renaissance to Early Industrialisation* (Amsterdam: Koninklijke Nederlandse Akademie van Wetenschappen, 2007), pp. 325–47.

56. Schama, *Citizens*, p. 131.
57. *L'Année littéraire*, 6 (1783), p. 137.
58. Sources cite different numbers for this famous launch although most agree that a sizable portion of the population in and around Paris attended the flight. See, for example, *Courrier d'Avignon*, 12 December 1783, p. 395.
59. *Gentleman's Magazine*, 58 (1785), p. 653.
60. *Centinel of Freedom*, 6 November 1810, [p. 3].
61. 'Expérience de la Muette', in F. de Saint-Fond (ed.), *Description des expériences aérostatiques*, vol. 2, p. 17.
62. *Affiches de Province*, 20 September 1783, p. 1086.
63. 'Expérience faite à Versailles', in Saint-Fond (ed.), *Description des expériences aérostatiques*, vol. 1, pp. 39–40.
64. 'Expérience faite à Paris au Champ de Mars', in de Saint-Fond (ed.), *Description des expériences aérostatiques*, vol. 1, p. 19.
65. J. E. Hodgson, *The History of Aeronautics in Great Britain* (London: Oxford University Press, 1924), p. 144.
66. J. Sadler, *Balloon: An Authentic Account of the Aerial Voyage of Messrs. Sadler and Clayfield, Who Ascended in a Most Magnificent Balloon From a Field in the Neighbourhood of Stoke's-Croft, Bristol, On Monday, September 24, 1810*, 2nd edn (Bristol, 1810), pp. 4–5.
67. *Massachusetts Centinel*, 30 March 1785, [p. 3].
68. *Independent Gazetteer*, 26 June 1784, [p. 2].
69. *Independent Ledger*, 6 December 1784, supplement, [p. 1].
70. A. F. Cradock, *Journal de Madame Craddock: Voyage en France (1783–1786)*, trans. and ed. O. Delphin Balleyguier (Paris: Perrin, 1896), pp. 1, 10.
71. Letter written by Prince Bariatinsky cited in N. Galitzyne, 'Les premières expériences de Montgolfier: D'après des documents russes', in *Annales internationals d'histoire. Congrès de Paris, 1900*, 7 vols in 2 tomes (Nendeln/Liechtenstein: Kraus Reprint, 1972), tome 2, vol. 5, pp. 146–54, 148–9.
72. L.S. Mercier in Mercier and R. de la Bretonne, *Paris le jour, Paris la nuit: 'Tableau de Paris', 'Le nouveau Paris' de Louis Sébastien Mercier & 'Les nuits de Paris' de Restif de la Bretonne*, ed. M. Delon and D. Baruch (Paris: R. Laffont, 1990.), p. 312.
73. J. Riskin, *Science in the Age of Sensibility: The Sentimental Empiricists of the French Enlightenment* (Chicago, IL: University of Chicago Press, 2002), cf. A. Vila, *Enlightenment and Pathology: Sensibility in the Literature and Medicine of Eighteenth-Century France* (Baltimore, MD: Johns Hopkins University Press, 1998).
74. C. Castelli, *Relation des expériences de la machine aérostatique de Don Paul Andréani* (n.p., 1784), pp. 20–1.
75. Ibid., pp. 21–2.
76. *Journal de Paris*, 16 July 1784, p. 847.
77. Pilâtre de Rozier to Faujas de Saint-Fond, 28 January 1784 in Faujas de Saint-Fond (ed.), *Description des expériences aérostatiques*, vol. 2, pp. 78–9.
78. *Précis historique de la grande expérience faite à Lyon la 19 Janvier 1784* ([1784]), pp. 14–15.
79. Marquis d'Arlandes, in Faujas de Saint-Fond (ed.), *Description des expériences aérostatiques*, vol. 2, p. 24.
80. 'Expérience faite à Paris au Champ de Mars', in Faujas de Saint-Fond (ed.), *Description des expériences aérostatiques*, vol. 1, p. 20.

81. Ibid., vol. 1, pp. 17–18.
82. *Courrier d'Avignon*, 5 December 1783, p. 388.
83. *Columbian Centinel*, 7 November 1795, [p. 3].
84. *Affiches de la Province de Bretagne* (1784), p. 227.
85. *Columbian Centinel*, 9 October 1790, p. 30.
86. Uncited references in Schama, *Citizens*, p. 130.
87. Baldwin, *Airopaidia*, p. 29.
88. Ibid., p. 37.
89. *Massachusetts Centinel*, 3 July 1784, [p. 1].
90. *Pennsylvania Packet*, 12 June 1784, [p. 2].
91. V. Lunardi, *An Account of the First Aerial Voyage in England* (London, 1784), p. 39.
92. *New Hampshire Spy*, 2 August 1788, p. 113.
93. *General Advertiser*, 18 November 1791, p. 2.
94. *L'Année littéraire*, 4 (1786), p. 308.
95. Sadler, *Balloon: An Authentic Narrative of the Aerial Voyage of Mr. Sadler, Across the Irish Channel* (Dublin, 1812), p. 4.
96. *New York Packet*, 23 September 1784, [p. 2].
97. *Columbian Centinel*, 16 January 1793, [p. 2].
98. Jeffries, *A Narrative of Two Aerial Voyages*, p. 25; M. B. Norton, 'America's First Aeronaut: Dr. John Jeffries', *History Today* 18 (1968), p. 725.
99. *New York Packet*, 21 February 1785, [p. 2].
100. Ibid., 21 February 1785, [p. 2], emphasis in original.
101. *Columbian Herald*, 30 November 1784, [p. 2].
102. *Morning Post*, 21 August 1784, n.p.
103. 'Extrait du Discours de M. Pilâtre de Rozier', in Faujas de Saint-Fond (ed.), *Description des expériences aérostatiques*, vol. 2, pp. 354–5.
104. Rolt, *The Aeronauts*, pp. 57–8.
105. Baldwin, *Airopaidia*, p. 159.
106. Mason, *Aeronautica*, pp. 276–7; *Morning Herald and Daily Advertiser*, 5 May 1785, n.p.
107. *Morning Herald and Daily Advertiser*, 29 June 1785, n.p.; *Gentleman's Magazine*, June 1785, p. 480; E. Rigby, *An Account of Mr. James Deeker's Two Aerial Expeditions From the City of Norwich* (Norwich: John Crouse, 1785), p. 15.
108. *Journal de Paris*, supplement, 13 July 1790, 4; *Reimpression de L'Ancien Moniteur*, 13 July 1790, p. 112.
109. J.-P. Blanchard, *Liste chronologique des ascensions de L'Aéronaute Blanchard* (n.p., n.d.).
110. Garnerin, *Parc de Mousseaux*.
111. *Boston Daily Advertiser*, 29 October 1816, [p. 2].
112. Rolt, *The Aeronauts*, pp. 73–4.
113. *Morning Herald and Daily Advertiser*, 14 October 1785, n.p.
114. Ibid., 24 March 1785, n.p.
115. *New York Packet*, 23 February 1786, [p. 1].
116. Blanchard, *Liste chronologique*.
117. *London Chronicle*, 16–18 September 1784, n.p.
118. *Otsego Herald*, 29 November 1798, [p. 2].
119. *Courrier d'Avignon*, 12 December 1783, p. 395.

120. Miss Port to her Father, 12 February 1785, in M. Delany, *The Autobiography and Correspondence of Mrs. Delany*, ed. S. Chauncey Woolsey (Boston, MA: Roberts Brothers, 1879), p. 430.
121. *Political Magazine*, September 1784, p. 235.
122. *Mr. Blanchard and a Lady* (1785).
123. *Morning Chronicle*, 21 September 1804, [p. 3].
124. Freneau, *The Poems of Philip Freneau*, p. 143.
125. *Morning Post*, 21 September 1784, n.p.
126. Ibid., 26 March 1785, n.p.
127. *Coquette Phisicienne*, n.d
128. *New York Packet*, 26 September 1785, [p. 4]; reprinted in *Essex Journal*, 26 October 1785, [p. 4].
129. 'Les Gaz', in Raoul de Cazenove, 'Ascension du Ballon le 'Gustave' à Lyon, 4 Juin 1784', *Revue Lyonnaise* 7 (1884), pp. 549–69, p. 558.
130. *Morning Herald and Daily Advertiser*, 12 November 1784, n.p. 'I am Lunardi – take guard of your hearts!'
131. 'On the Lunardi Bonnet and Garters Worm by the Ladies', in *An Asylum for Fugitive Pieces in Prose and Verse*, 4 vols (London: Debrett, 1785–93), vol. 1, p. 55.
132. *Morning Post*, 8 October 1784, n.p.
133. *New York Packet*, 20 January 1785, [p. 2].
134. *Morning Post*, 23 September 1784, n.p.; emphasis in original.
135. *Morning Herald*, 2 January 1786, [p. 2]; emphasis in original.
136. On the Beggar's Benison see D. Stevenson, *The Beggar's Benison: Sex Clubs of Enlightenment Scotland and their Rituals* (East Linton: Tuckwell Press, 2001), pp. 161–2.
137. Lunardi, *An Account of Five Aerial Voyages*, p. 38.
138. *Morning Post*, 28 October 1784, n.p.
139. *Pennsylvania Packet*, 26 February 1784, [p. 3].
140. Uncited quote in Norton, 'America's First Aeronaut', 724.
141. *Walker's Hibernian Magazine*, October 1785, p. 507.
142. *Publick Ledger*, 15 January 1784, n.p.
143. *Oracle of Dauphin*, 1 November 1802, [p. 1].
144. *Gentleman's Magazine*, September 1784, p. 711.
145. Ibid., p. 711. On curiosity in general as well as the Bottle Conjuror, see B. Benedict, *Curiosity: A Cultural History of Early Modern Inquiry* (Chicago, IL: University of Chicago Press, 2001).
146. *Massachusetts Centinel*, 24 November 1784, [p. 1].
147. *New York Packet*, 21 July 1785, [p. 2].
148. Bachaumont quoted in Rolt, *The Aeronauts*, pp. 28–9.
149. For Argand and Lunardi see Rolt, *The Aeronauts*, pp. 62 and 70–71.
150. *Gentleman's Magazine*, August 1814, pp. 180–1.
151. *Morning Herald*, 26 December 1783, n.p.
152. *Petit Journal du Palais Royal, ou Affiches, annonces et avis divers* (Paris: Caveau, 1789), p. 28.
153. *Courrier d'Avignon*, 10 October 1783, p. 324.
154. Watt to Black, 25 September 1783 in *Partners in Science: Letters of James Watt and Joseph Black*, ed. E. Robinson and D. McKie (Cambridge, MA: Harvard University Press, 1970), p. 128.
155. *Pennsylvania Packet*, 15 November 1783, [p. 2].

156. Ibid., 25 December 1783, [p. 2].
157. Blanchard quoted in *Political Magazine*, 1784, p. 177.
158. Rolt, *The Aeronauts*, p. 76. On Lunardi in Spain see Rocamora, *Historia de la navegación aérea en Barcelona*, pp. 23–34.
159. *New York Packet*, 16 February 1786, [p. 2].
160. *Courier de l'Europe*, 9 September 1783; *Journal Encyclopédique*, October 1783, p. 126; A. Rivarol, 'Lettre à Monsieur le Président de ***. Sur le Globe Airostatique, sur les Têtes parlantes, & sur l'état présent de l'opinion publique à Paris', in *Oeuvres*, 5 vols (Paris, 1808; rpt. Geneva: Slatkine, 1968), vol. 2, p. 216; 'Lettre au Rédacteur de *l'Année Littéraire*, sur l'exprérience de la Machine Aérostatique', *L'Année littéraire*, 6 (1783), p. 140.
161. Lunardi, *An Account of the First Aerial Voyage*, 37; Rolt, *The Aeronauts*, p. 71.
162. *Pennsylvania Packet*, 18 September 1784, [p. 2].
163. *Commercial Advertiser*, 7 May 1800, [p. 3].
164. *Journal de Paris*, 8 June 1784, p. 328.
165. *Morning Post*, 11 August 1784, n.p.
166. Lunardi, *An Account of Five Aerial Voyages*, p. 93.
167. *Courrier d'Avignon*, 30 December 1783, p. 416.
168. *National Advocate*, 19 August, 1817, [p. 2]; *Daily National Intelligencer*, 12 August 1816, [p. 3]; *Carlisle Gazette*, 30 January 1793, [p. 3].
169. *Boston Gazette*, 6 September 1802, [p. 4].
170. *Procès-Verbal trés-interessant du voyage aerien* (Paris, 1791), p. 10.
171. *Journal de Nancy*, 12:18 (1783), p. 96. The peasant says that he 'does not know balloons' but 'he saw a large bird'.
172. *Journal de Paris*, 29 September 1784, pp. 1132–3.
173. *Journal Encyclopédique*, September 1785, pp. 326–27; *Journal de Paris*, 18 August 1785, 952; *Morning Herald and Daily Advertiser*, 10 August 1785, n.p; *Pennsylvania Packet*, 15 November 1785, [p. 2].
174. Rolt, *The Aeronauts*, p. 62.
175. *Gentleman's Magazine*, October 1784, p. 791.
176. For more on riots, see Chapter 5.
177. *Independent Ledger*, 28 March 1785, [p. 3].
178. Baldwin, *Airopaidia*, p. 24.
179. *Morning Post*, 21 September 1784, n.p.
180. *Gentleman's Magazine*, March 1784, p. 228.
181. *Annales politiques, civiles et littéraires du dix-huitième siècle* ed. S.-N.-H. Linguet, 19 vols (London: 1777–1792), vol. 10, p. 365.
182. Gimbel Collection, XL15 1565.
183. *Freeman's Journal*, 14 December 1785, [p. 2].
184. *Salem Gazette*, 16 September 1796, [p. 1].
185. *Courrier d'Avignon*, 1 June 1784, pp. 179–80.
186. B. Franklin, *Benjamin Franklin on Balloons: A Letter Written from Passy, France, January Sixteenth MDCCLXXIV* (Saint Louis: W.K. Bixby, 1922), n.p.
187. *The General Evening Post*, 3–5 May 1785, [p. 1].
188. Lecornu, *La navigation aérienne*, p. 109.
189. Faujas de Saint-Fond (ed.), *Description des expériences aérostatiques*, vol. 2, p. 40. Many commentators mentioned the use of guards at launches: see, for example, Rivarol, 'Lettre à Monsieur le Président de ***'; and J.-G. Wille, *Mémoires et journal de J.-G. Wille*,

graveur du roi, ed. Georges Duplessis, 2 vols (Paris: Renouard, 1857), vol. 1, pp. 73–4. I discuss the use of guards to maintain order in more detail below.

190. *City of Washington Gazette*, 1 August 1818, [p. 2].
191. Lunardi, *An Account of Five Aerial Voyages*, p. 29; for more on his earlier fears at the Artillery Ground launch see Lunardi, *An Account of the First Aerial Voyage*, pp. 26–7, 28.
192. *City of Washington Gazette*, 14 September 1819, [p. 3]. This riot was also reported in the *Portsmouth Oracle*, 18 September 1819, [p. 2]; *Daily National Intelligencer*, 14 September 1819, [p. 2]; and *Farmer's Cabinet*, 18 September 1819, [p. 3].
193. This point of view will be discussed further in Chapter 5.
194. *Morning Post*, 21 August 1784, n.p.
195. Ibid., 14 August 1784, n.p.
196. Cf. V. Turner, *From Ritual to Theater: The Human Seriousness of Play* (New York: Performing Arts Journal Publications, 1982).

4 Controlling the Skies: States and Balloons

1. J.-F. Depuis-Delcourt, *Des Ballons dans les fêtes publiques* (Imprimerie Schneider et Langrand, 1846), p. 9; *New-England Palladium*, 29 October 1805, [p. 2].
2. Depuis-Delcourt, *Des Ballons dans les fêtes publiques*, pp. 9–10.
3. Godechot, 'L'Aérostation militaire sous le directoire', p. 225.
4. 'Rapport du bureau contral du 11 Thermidor', in F.-A. Aulard (ed.), *Paris pendant la reaction Thermidorean et sous le Directoire*, 5 vols (Paris, 1898–1902), vol. 5, p. 15.
5. *Philadelphia Gazette*, 9 December 1802, [p. 3].
6. *Journal de Paris*, supplement, 24 December 1789, p. 2; ibid., supplement, 25 December 1789, p. 1.
7. *Philadelphia Gazette*, 9 September 1801, [p. 2]; *Alexandria Advertiser*, 12 September 1801, [p. 2].
8. *New York Daily Gazette*, 30 September 1790, p. 934.
9. *Fête nationale, Qui sera célébrée aujourd'hui au Champ-de-Mars, aux Champs-Elysées, à la Halle, & sur la place de la Bastille* ([Paris, *c.* 1790]), pp. 4–5.
10. *Fête aérostatique, qui sera célébrée aujourd'hui au Champs de Mars* (N.p., n.d.), p. 5. This pamphlet is very similar to the one titled *Fête nationale, Qui sera célébrée aujourd'hui au Champ-de-Mars, aux Champs-Elysées, à la Halle, & sur la place de la Bastille* ([Paris, *c.* 1790].
11. *Reimpression de l'Ancien Moniteur*, 32 vols (Paris: Plon, 1858–63), vol. 29, p. 368.
12. *Procès-Verbal trés-interessant du voyage aerien* (Paris, 1791), pp. 3–4.
13. B. Lallemand de Saint-Croix, *Procès-verbal très intéressant du voyage aérien qui a eu lieu aux Champs-Elysées le 18 septembre 1791, jour de la proclamation de la constitution* (Paris: De l'Imprimerie du patriote françois, 1791), pp. 3–5. Also see P. Oberling, 'A History of Turkish Aviation. Part I: Aerostation among the Ottomans', *Archivum Ottomanicum* 9 (1984), p. 136.
14. Garnerin, *Parc de Mousseaux*.
15. Anon. (ed.), *The Romance of Ballooning*, p. 82.
16. *Baltimore Patriot*, 4 March 1816, [p. 2].
17. The Prince de Ligne quoted in F. Freska (ed.), *A Peace Congress of Intrigue*, trans. Harry Hansen (New York: The Century Co., 1919), p. 81.
18. Freska, (ed.), *A Peace Congress of Intrigue*, p. 83.
19. *Morning Post*, 25 September 1785, n.p; *Bath Chronicle* 30 September 1785, p. 1.

20. *Morning Post*, 29 December 1784, n.p.
21. Lunardi, *An Account of the First Aerial Voyage in England*, p. 2.
22. L. S. Mercier, *Mon Bonnet de Nuit*, ed. J.-C. Bonnet (Paris: Mercure de France, 1999), pp. 560–61.
23. *Pennsylvania Packet*, 20 April 1784, [p. 2].
24. *Independent Ledger*, 13 December 1784, [p. 4].
25. Baldwin, *Airopaidia*, pp. 261–2.
26. *Massachusetts Centinel*, 20 November 1784, [p. 3]. In this article, the anonymous author claimed 'the British seem not to deserve the smallest credit, either in the invention, or even improvement of the Balloon'.
27. *Massachusetts Centinel*, 21 March 1787, p. 4.
28. Linguet (ed.), *Annales politiques, civiles et littéraires du dix-huitième siècle*, 19 vols (London: 1777–92), vol. 11, p. 37.
29. *Bath Chronicle*, 4 March 1784, p. 4.
30. *Madame Blubber's Last Shift, or the Aerostatic Dilly*, 29 April 1784 (Broadside).
31. *Weekly Oracle*, 17 December 1798, [p. 3].
32. *Federal Gazette*, 31 October 1797, [p. 2].
33. Brissot de Warville quoted in Venturi, *The End of the Old Regime in Europe*, vol. 2, p. 440.
34. *Morning Post*, 24 September 1784, n.p; and Ibid., 25 September 1784, n.p.
35. Ibid., 1 November 1784, n.p.
36. *Providence Gazette*, 13 January 1798, [p. 2].
37. *Morning Post*, 16 October 1784, n.p.
38. Ibid., 16 October 1784, n.p
39. *Massachusetts Gazette*, 13 March 1787, [p. 2].
40. *Weekly Messenger*, 9 July 1818, [611].
41. *Boston Daily Advertiser*, 3 July 1818, [p. 2].
42. *Poulson's American Daily Advertiser*, 4 July 1809, [p. 3].
43. *Independent Chronicle*, 24 January 1793, [p. 3]; *American Apollo*, 25 January 1793, [p. 3].
44. *United States Chronicle*, 4 March 1802, [p. 2].
45. *Boston Daily Advertiser*, 27 January 1814, [p. 2].
46. *Post-Boy*, 8 July 1806, p. 215.
47. *Weekly Wanderer*, 17 November 1806, [p. 4].
48. Venturi, *The End of the Old Regime*, vol. 2, p. 935.
49. *City Gazette*, 18 February 1803, [p. 2].
50. Ibid.
51. *Columbia Centinel*, 2 January 1799, [p. 2].
52. *Relation du voyage aérostatique du Général Buonaparte* (N.p., n.d.), p. 2.
53. *The Mail, or Claypoole's Daily Advertiser*, 1 October 1792, [p. 2].
54. For a positive assessment see Money, *A Short Treatise on the Use of Balloons and Field Observations in Military Operations*; and *L'art de la guerre changé par l'usage des machines aérostatiques* (N.p., 1784).
55. Horace Walpole quoted in L.T.C. Rolt, *The Aeronauts*, p. 159.
56. Gillispie, *The Montgolfier Brothers*, p. 125.
57. Catherine the Great quoted in Alexander, 'Aeromania', p. 503.
58. Franklin quoted in Connaughton, 'Ballomania', p. 72.
59. *Affiches de Paris*, 26 October 1783, p. 1231.
60. *Courrier d'Avignon*, 7 November 1783, p. 356.

61. *Daily Advertiser*, 14 November 1796, [p. 2].
62. *Reimpression de l'Ancien Moniteur*, 32 vols (Paris: Plon, 1858–63), vol. 29, p. 73.
63. For the United States see, for example, the *Green Mountain Patriot*, 16 March 1798, [p. 3]; *Courier of New Hampshire*, 13 March 1798, [p. 2]; and *Oracle of the Day*, 3 March 1798, [p. 2].
64. For Blanchard see the *New-York Gazette*, 21 March 1798, [p. 3]; for Garnerin as the Field Marshall of the Balloon Army see *Connecticut Herald*, 21 June 1808, [p. 4]; for nonsense see *Federal Gazette*, 12 March 1798, [p. 3].
65. For 1793 see the *Middlesex Gazette*, 29 June 1793, [p. 2]. For the later reference see the *City Gazette*, 26 September 1799, [p. 2].
66. *Morning Chronicle*, 10 September 1803, [p. 3]; reprinted in *Gazetteer*, 28 September 1803, [p. 4].
67. *Morning Chronicle*, 16 February 1804, [p. 3]; this rumour is repeated in the *Morning Chronicle*, 25 May 1804, [p. 3].
68. *Essex Register*, 26 March 1808, [p. 3].
69. *Windham Herald*, 17 December 1812, n.p.
70. On the Compagnie des Aérostiers see Gillispie, *Science and Polity in France*, pp. 371–3; P. Bret, 'Recherche scientifique, innovation technique et conception tactique d'une arme nouvelle: l'aérostation militaire (1793–1799)', in J.-P. Charnay, et al. (eds), *Lazare Carnot, ou le savant-citoyen* (Paris: Presses de l'Université de Paris, 1990), pp. 429–51; J. Godechot, 'Les premiers soldats du ciel: Débuts de l'aérostation militaire', *Historama*, 11 (1985), pp. 24–8; ibid., 'L'Aérostation militaire sous le directoire', *Annales historiques de la Révolution française* 8 (1931), pp. 213–28; L. Robineau, 'Lazare Carnot et les Compagnies d'aérostiers', *Revue historique des Armées*, 2 (1989), pp. 101–10; and D. Lacroix, *Les aérostiers militaires du Chateau de Meudon (1794–1884)* (Paris: Auguste Ghio, 1885). More generally see J. Duhem, *Histoire de l'arme aérienne avant le moteur* (Paris: Nouvelles editions latines, 1964).
71. For Coutelle see J. M. J. Coutelle, *Sur l'aérostat employé aux armées de Sambre-et-Meuse et du Rhin* (Paris: Crapelet, 1829). On Conté see P. Bret, 'Nicolas-Jacques Conté ou l'intelligence technique', *La Revue: Musée des arts et métiers*, 25 (December 1998), pp. 4–13; A. Queruel, *Nicolas-Jacques Conté (1755–1805), un inventeur de génie: Des crayons à l'Expédition d'Egypte en passant par l'aérostation militaire* (Paris: L'Harmattan, 2004); and F. Cornu-Thenard, *Nicolas-Jacques Conté: Curieux autodidacte du XVIIIe siècle* (Dijon: Jobard, 1955).
72. Gillispie, *Science and Polity in France*, p. 373; Shelby McCloy, *French Inventions of the Eighteenth Century* (Louisville, KY: University of Kentucky Press, 1952), p. 23.
73. *City Gazette*, 20 August 1794, [p. 2].
74. *Middlesex Gazette*, 4 October 1794, [p. 3].
75. Baron de Selle de Beauchamp, *Souvenirs de la fin du dix-huitième siècle, extraits des mémoires d'un officier des aérostiers aux armies de 1793–1799* (Paris: Ledoyen et Paul Giret, 1853), p. 40; also see Godechot, 'L'Aérostation militaire sous le directoire', p. 216.
76. E. de Villiers du Terrage, *Journal et souvenirs sur l'expédition d'Egypte (1798–1801)* (Paris: Plon, 1899), pp. 86–7.
77. J. Langins, 'Hydrogen Production for Ballooning during the French Revolution: An Early Example of Chemical Process Development', *Annals of Science*, 40 (1983), p. 555; Godechot, 'L'Aérostation militaire sous le directoire', p. 223.
78. Faujas de Saint Fond (ed.), *Description des expériences aérostatiques*, vol. 2, p. 12.

79. Schofield, *The Lunar Society of Birmingham*, p. 252.
80. *Columbian Centinel*, 29 October 1796, [p. 1].
81. *Morning Post*, 19 March 1785.
82. *Journal de Paris*, cited in Rolt, *The Aeronauts*, p. 43.
83. R. Fox, 'Science and Government', in *The Cambridge History of Science*, vol. 4., *Eighteenth-Century Science*, ed. R. Porter (Cambridge: Cambridge University Press, 2003), pp. 107–28.
84. Glines, 'First in America's Skies' *Aviation History*, p. 28; Blanchard, *Journal of My Forty-Fifth Ascension*.
85. Rolt, *The Aeronauts*, pp. 34–5.
86. Hardy, cited in J. M. Hunn, 'The Balloon Craze in France, 1783–1799: A Study in Popular Science' (Ph.D. dissertation, Vanderbilt University, 1982), p. 131.
87. Rivarol, 'Lettre à Monsieur le Président de ***'. Also see Galitzyne, 'Les premières expériences de Montgolfier', tome II, vol. 5, p. 149; and *Courier de l'Europe*, 9 September 1783, p. 167.
88. *Avertissement au Peuple, sur l'Enlevement des Ballons ou Globes en l'air* (Paris: Herissant, 1783).
89. *Ordonnance de Police, qui fait defenses de fabriquer & faire enlever des Ballons* (Paris: Pierres, 1784). This information was published and displayed all over Paris and was also repeated by town criers. It was also reprinted in a variety of newspapers such as the *Courrier d'Avignon*, 18 May 1784, p. 163.
90. *Gentleman's Magazine*, 55 (1784), p. 433.
91. *Ordonnance de Police, qui fait défenses à toutes personnes de quelque qualité & condition qu'elles soient, de fabriquer & faire enlever aucuns ballons* (Angers: Mame, 1784). This information also appeared in newspapers. See, for example, *Affiches d'Angers*, 4 June 1784, p. 100.
92. For the limitations on who could launch balloons see the *Paris Ordonnance de Police*, pp. 1–2, and the *Ordonnance de Police* (Angers).
93. Lantier, 'Retombées aérostatiques sur la généralité de Caen', p. 272.
94. Ibid., p. 272.
95. Ibid., p. 274.
96. Ibid., p. 276.
97. Holl, 'Les débuts de l'aviation à Strasbourg'.
98. P. Verhagen, 'Les premiers ballons en Belgique, 1783–1786', *Annales de la Société royale d'archéologie de Bruxelles*, 13 (1899), p. 370.
99. *Monthly Review*, May 1784, p. 408.
100. *Essex Journal*, 3 September 1784, [p. 3].
101. *Independent Journal*, 10 February 1787, [p. 2].
102. *American Recorder*, 6 April 1787, [p. 1].
103. Prussian and Milanese examples from the Russian newspaper *Moskovskie vedomosti*, translated by and cited in Alexander, 'Aeromania', p. 511.
104. Ibid., p. 512.
105. *Ordonnantie van den Keyser, Raekende de Ballons ofte Locht-Tuygen met Vuer, genoemt Montgolfieres* (Brussels: Pauwels, 1786).
106. *Pennsylvania Evening Herald*, 5 October 1785, p. 84.
107. *Pennsylvania Packet*, 5 January 1786, [p. 2].
108. Ibid., 1 January 1785, [p. 2].
109. Alexander, 'Aeromania', pp. 514–15.

110. Rolt, *The Aeronauts*, p. 46.
111. Schama, *Citizens*, p. 131.
112. Linguet (ed.), *Annales politiques*, vol. 10, p. 365.
113. *Mémoires secrets pour servir à l'histoire de la République des Lettres en France depuis 1762 jusqu'à nos jours*, 36 vols (London: John Adamson, 1784–1789), vol. 23 (12 October 1783), p. 239.
114. *Morning Post*, 21 September 1784, [p. 3].
115. Ibid. 13 March 1784, [p. 3].
116. *Gentleman's Magazine*, 54 (March 1784), p. 228.
117. Henry Laurens to Benjamin Franklin, 28 November 1783, in *The Papers of Henry Laurens*, Philip M. Hamer, ed. (Columbia: University of South Carolina Press, 2003).
118. Rolt, *The Aeronauts*, p. 69.
119. For the *London Times* and its comments see the citation in M. J. Quinlan, 'Balloons and the Awareness of a New Age', *Studies in Burke and His Time*, 14 (1972–3), p. 227.
120. *Publick Advertiser*, 27 September 1784, n.p.; *General Advertiser*, 9 September 1784, n.p.
121. *Freeman's Journal*, 6–8 January 1785, n.p.
122. *Commercial Advertiser*, 8 October 1816, [p. 2].
123. On thieves in France see Linguet (ed.), *Annales politiques*, vol. 10, p. 365.
124. See R. Darnton, *Mesmerism and the End of the Enlightenment in France* (Cambridge, MA: Harvard University Press, 1968), pp. 23–4.
125. *Independent Chronicle*, 1 September 1785, [p. 4].
126. Franklin, *Benjamin Franklin on Balloons*, n.p.
127. *Morning Herald and Daily Advertiser*, 12 August 1784, [p. 1]. On Moret see the brief synopsis in Gardiner, *Man in the Clouds*, pp. 35–6.
128. See E. P. Thompson, 'The Moral Economy of the English Crowd in the Eighteenth Century', *Past and Present*, 50 (1971), pp. 76–136.
129. *Political Magazine*, August 1784, p. 88.
130. Ibid., September 1784, p. 232.
131. V. Lunardi, *An Account of the First Aerial Voyage in England* (London, 1784), pp. 26–7.
132. Lunardi, *An Account of the First Aerial Voyage*, p. 28.
133. Lunardi, *An Account of Five Aerial Voyages in Scotland*, p. 29 (emphasis in original).
134. Rolt, *The Aeronauts*, pp. 75–6.
135. *Gazetteer and New Daily Advertiser*, 3 January 1785, n.p.
136. *Affiches de Province*, 14 February 1784, pp. 94–5; *Journal de Paris*, 26 February 1784, pp. 257–8.
137. *Journal de Paris*, 21 March 1784, p. 361; *Affiches de Province*, 10 July 1784, p. 395; *Affiches de Paris*, 1 April 1784, p. 858.
138. For his public lecture courses on physics see, for example, *Journal de Paris*, 4 December 1783, p. 1388; and *Affiches de Paris*, 21 April 1783, p. 963. For Miollan's connection with Marat see N. B. Mandelbaum, 'Jean-Paul Marat: The Rebel as Savant (1743–1788). A Case Study in Careers and Ideas at the End of the Enlightenment', (Ph.D. Dissertation, Columbia University 1977), p. 505.
139. Gillispie, *The Montgolfier Brothers*, p. 97.
140. Cradock, *Journal de Madame Craddock*, pp. 62–3.
141. N. M. Karamzin, *Letters of a Russian Traveler, 1789–1790*, trans. Florence Jonas (New York: Columbia University Press, 1957), pp. 312–14.
142. *Norwich Packet*, 7 October 1784, [p. 1].
143. Meaudre de Lapouyade, *Les premiers aeronauts bordelais*, pp. 28–30

144. *The General Evening Post*, 12 July–14 July 1785, p. 4. See Chapter 2 for more details.
145. *The Mail*, 23 August 1791, p. 2.
146. *Morning Post*, 21 August 1784, n.p.
147. *New York Journal*, 19 August 1784, [p. 3].
148. *New Jersey Journal*, 19 July 1786, [p. 2].
149. Lecornu, *La navigation aérienne*, p. 109.
150. *Gentleman's Magazine*, August 1802, p. 776.
151. *Farmer's Cabinet*, 18 September 1819, [p. 3]; see Michel's account in the *City of Washington Gazette*, 14 September 1819, [p. 3]. Other accounts appeared in the *Portsmouth Oracle*, 18 September 1819, [p. 2]; and the *Daily National Intelligencer*, 14 September 1819, [p. 2].
152. *Gentleman's Magazine*, September 1784, p. 64.
153. *The General Evening Post*, 3–5 May 1785, [p. 1].
154. *Aris's Birmingham Gazette*, 3 January 1785, n.p.
155. *Columbia Centinel*, 6 November 1790, p. 63.
156. Thompson, 'The Moral Economy of the English Crowd', pp. 76–136; and Turner, *From Ritual to Theater*.
157. Baldwin, *Airopaidia*, p. iii.
158. J. Sadler, *Balloon: An Authentic Narrative of the Aerial Voyage of Mr. Sadler, Across the Irish Channel* (Dublin, 1812), p. 6.
159. Ibid., p. 7.
160. Sadler, *Balloon: An Authentic Narrative of the Aerial Voyage of Mr. Sadler, Across the Irish Channel*, p. 3.
161. Cited in Anon. (ed.), *The Romance of Ballooning*, p. 80.
162. S. Agin, 'Le public des ascensions en montgolfière et la democratization de la science au XVIIIe siècle', in *Cahiers de littérature française*, M. Delon and J. M. Goulemot (eds), vol. 5, *Ballons et regards d'en haut* (Paris: L'Harmattan; Bergame: Bergamo University Press, 2007), pp. 17–18. Also see Fox, 'Science and Government', p. 108; cf. L. P. Williams, 'The Politics of Science in the French Revolution', in M. Clagett (ed.), *Critical Problems in the History of Science* (Madison, WI: The University of Wisconsin Press, 1959), pp. 291–308; he notes that during the French Revolution, Newtonian science was viewed by some as bad because it was too elitist.
163. Fox, 'Science and Government', pp. 112–13.

5 Consuming Balloons

1. *Morning Post*, 19 March 1785, p. 1.
2. Rolt, *The Aeronauts*, p. 82.
3. There are several good studies exploring consumption during the earlier period. See, for example, C. Mukerji, *From Graven Images: Patterns of Modern Materialism* (New York: Columbia University Press, 1983); and E. Welch, *Shopping in the Renaissance: Consumer Cultures in Italy, 1400–1600* (New Haven, CT: Yale University Press, 2005). In general see de Vries, *The Industrious Revolution*.
4. See, for example, Mukerji, *From Graven Images*; Baudrillard, *The Consumer Society*; and M. Douglas and B. Isherwood, *The World of Goods: Towards an Anthropology of Consumption* (London: Routledge, 1996).
5. *Publick Advertiser*, 12 April 1785, n.p.

6. For an introduction to consumption in early modern Europe see N. McKendrick, J. Brewer and J. H. Plumb, *The Birth of a Consumer Society: The Commercialization of Eighteenth-Century England* (Bloomington, IN: Indiana University Press, 1982). For a critique of McKendrick see many of the essays in J. Brewer and R. Porter (eds), *Consumption and the World of Goods* (London: Routledge, 1993). For a more general survey of the ways scholars have approached this topic see, for example, D. Miller (ed.), *Acknowledging Consumption: A Review of New Studies* (London: Routledge, 1995).
7. On novelty see M. Bianchi, 'Taste for Novelty and Novel Tastes: The Role of Human Agency in Consumption', in *The Active Consumer: Novelty and Surprise in Consumer Choice*, ed. M. Bianchi (London: Routledge, 1998), pp. 64–86. Also see G. McCracken, *Culture and Consumption: New Approaches to the Symbolic Character of Consumer Goods and Activities* (Bloomington, IN: Indiana University Press, 1990), esp. ch. 1; and S. Schaffer (who is citing J. H. Plumb), 'The Consuming Flame: Electrical Showmen and Tory Mystics in the World of Goods', in Brewer and Porter (eds), *Consumption and the World of Goods*, pp. 489–90.
8. Paul Glennie has noted that most historians of early modern Europe analyse consumer culture through material culture rather than cultural representations (like a launch) because of the accessibility of sources. See Glennie, 'Consumption within Historical Studies', in *Acknowledging Consumption*, p. 166.
9. C. Campbell, 'Sociology of Consumption', in *Acknowledging Consumption*, p. 111.
10. Glennie, 'Consumption within Historical Studies', in *Acknowledging Consumption*, p. 180.
11. See D. Miller, *Material Culture and Mass Consumption* (London: Blackwell, 1987), p. 151; and, more generally, P. Bourdieu, *Distinction: A Social Critique of the Judgement of Taste*, trans. R. Nice (Cambridge, MA: Harvard University Press, 1984); and idem., *Outline of a Theory of Practice*, trans. R. Nice (Cambridge: Cambridge University Press, 1977).
12. V. R. Schwartz, *Spectacular Realities: Early Mass Culture in Fin-de-Siècle Paris* (Berkeley, CA: University of California Press, 1998).
13. *Morning Herald and Daily Advertiser*, 20 January 1785.
14. *Chronique de Paris*, 25 December 1789, p. 496 (Christmas festival); *Morning Herald and Daily Advertiser*, 15 November 1784 (party atmosphere).
15. *Affiches de Toulouse et du Haut-Languedoc*, 4 February 1784, p. 21 and *L'Année littéraire*, p. 142 (1785), p. 142 (spectacle); *Gazetteer and New Daily Advertiser*, 17 August 1784 and *Morning Herald and Daily Advertiser*, 5 May 1785 (curiosity).
16. *Morning Post*, 2 October 1784. On the connection with luxury see M. Kwass, 'Ordering the World of Goods: Consumer Revolution and the Classification of Objects in Eighteenth-Century France', *Representations*, 82 (2003), pp. 87–116; and M. Berg and E. Eger (eds), *Luxury in the Eighteenth-Century: Debates, Desires and Delectable Goods* (New York: Palgrave, 2003).
17. For launches as experiments see, for example, *Freeman's Journal*, 23–5 June 1785; and *Affiches de Paris*, 3 October 1790, p. 3066.
18. *Journal de Paris*, 13 June 1786, p. 677 (meteorological equipment); Lunardi, *An Account of the First Aerial Voyage in England* (London, 1784), pp. 8–9; *Morning Post*, 16 January 1784.
19. *Morning Herald and Daily Advertiser*, 3 June 1785; and *Morning Post*, 12 May 1785.
20. On the utility of balloons see, for example, *Gazetteer and New Daily Advertiser*, p. 30 April 1784; and *Morning Herald and Daily Advertiser*, 12 August 1784.

21. Shapin, 'The Audience for Science in Eighteenth-Century Edinburgh', p. 95.
22. On the efforts of the Montgolfier brothers to gain state patronage, see C. Gillispie, *The Montgolfier Brothers*, passim.
23. Faujas de Saint-Fond (ed.), *Description des expériences aérostatiques*, vol. 1, pp. 8–9.
24. On Lavoisier see M. G. Kim, '"Public" Science: Hydrogen Balloons and Lavoisier's Decomposition of Water', *Annals of Science*, 63 (2006), pp. 291–318. Kim argues that Lavoisier's two experiments, one in public and the other for a select group of savants, indicates that the Académie Royale des Sciences was fully involved in ballooning and using the new science to renegotiate their authority over the public's view of science; however, in doing so Kim conflates Lavoisier's chemical successes with an Academy that had really had no control over the practice of aeronautics.
25. On the launch in Rodez, see *Affiches de Toulouse et du Haut-Languedoc*, 18 August 1784, p. 136.
26. For the subscription goals see Mathon de la Cour, 'Lettre de Mathon de la Cour, Directeur de l'académie des sciences de Lyon, sur l'expérience de l'Aérostate que M. de Montgolfier a fait élever à Lyon', in Faujas de Saint-Fond (ed.), *Description des experiences aérostatiques*, vol. 2, pp. 84–5. For a list of the subscribers see the 'Etat des souscriptions pour la grande experience du ballon aérostatique', in de Cazenove, *Premiers voyages aériens à Lyon en 1784*, pp. 60–3. On the social status of the subscribers and the satire of Mlle Salade see Hunn, 'The Balloon Craze in France', pp. 219–20.
27. *Affiches de Province*, 3 November 1783, pp. 1264–5.
28. Ibid., 20 March 1784, pp. 171–2.
29. *Courrier d'Avignon*, 9 April 1784, pp. 119–20. On Charles and the Robert brothers tactics see Gillispie, *The Montgolfier Brothers*, p. 57. On the Robert brothers see Robert and Robert, *Mémoire sur les Expériences Aérostatiques*. On the success of their subscription see Hunn, 'The Balloon Craze in France', pp. 136–7. Hunn cites the commentator Simeon-Prosper Hardy, *Mes loisirs*, 8 vols, Bibliothèque Nationale, Ancien fond français, 6680–7, citation in 6684, p. 390.
30. *Morning Post*, 19 March 1785.
31. See, for example, Schaffer, 'The Consuming Flame'; J. Secord, 'Newton in the Nursery: Tom Telescope and the Philosophy of Tops and Balls, 1761–1838', *History of Science*, 23 (1985), pp. 127–51; A. Walters, 'Conversation Pieces: Science and Politeness in Eighteenth-Century England', *History of Science*, 35 (1997), pp. 121–54; idem., 'Ephemeral Events: English Broadsides of Early Eighteenth-Century Solar Eclipses', *History of Science*, 37 (1999), pp. 1–43; L. Stewart, *The Rise of Public Science: Rhetoric, Technology, and Natural Philosophy in Newtonian Britain, 1650–1750* (Cambridge: Cambridge University Press, 1992); Geoffrey Sutton, *Science for a Polite Society: Gender, Culture, & the Demonstration of Enlightenment* (Boulder, CO: Westview Press, 1995); L. E. Robbins, *Elephant Slaves and Pampered Parrots: Exotic Animals in Eighteenth-Century Paris* (Baltimore, MD: Johns Hopkins University Press, 2002); Lynn, *Popular Science and Public Opinion*.
32. J. Golinski, 'Barometers of Change: Meteorological Instruments as Machines of Enlightenment', in Clark, Golinski and Schaffer (eds), *The Sciences in Enlightened Europe*, pp. 69–93, pp. 78, 71.
33. In this case, an additional link with ballooning can be identified; subscribers to hospitals often received recognition through the publication of subscriptions lists or an even more prominent posting of a list of subscribers at the institution itself; this often took the form of a plaque on the building. This allowed subscribers' names to become known to the

citizens of the community and demonstrated, in a very public way, who had participated in the funding of a particular charitable institution. It could also be viewed as self-promotion to some extent as middling subscribers could associate their names with those of the more socially prestigious who had also participated in the project. Some of the early balloon subscriptions followed this practice thus giving participants the added honour of having their names associated with the launch.

34. On the *Encyclopédie* see J. Lough, *The Encyclopédie* (New York: David McKay, 1971), pp. 19–20; and R. Darnton, *The Business of Enlightenment: A Publishing History of the Encyclopédie, 1775–1800* (Cambridge, MA: The Belknap Press, 1979); for books more generally see F. Waquet, 'Book Subscriptions in Early Eighteenth-Century Italy', *Publishing History*, 33 (1993), pp. 77–88; for libraries see D. Allan, 'Eighteenth-Century Private Subscription Libraries and Provincial Urban Culture: The Amicable Society of Lancaster, 1769–*c.* 1820', *Library History*, 17 (2001), pp. 57–76; and R. Chartier, 'Urban Reading Practices, 1660–1780', in *The Cultural Uses of Print in Early Modern France*, trans. L. G. Cochrane (Princeton, NJ: Princeton University Press, 1987), pp. 183–239, esp. 202–15. For hospitals see A. Berry, '"Balancing the Books": Funding Provincial Hospitals in Eighteenth-Century England', *Accounting, Business and Financial History*, 7 (1997), pp. 1–30; for ships see Y. Aoki, 'To be a Member of the Leading Gentry: The Suffolk Voluntary Subscriptions of 1782', *Historical Research*, 76 (February 2003): 78–92; on Mozart see T. Christiensen, 'Concert', in A. C. Kors (ed.), *Encyclopedia of the Enlightenment*, 4 vols (Oxford: Oxford University Press, 2003), vol. 1, pp. 282–3. More generally see J. Brewer, 'Commercialization and Politics', and J. H. Plumb, 'The Commercialization of Leisure in Eighteenth-Century England', both in McKendrick (et al.), *The Birth of a Consumer Society*, pp. 224–35 and 282–3.
35. Jeffries, *A Narrative of Two Aerial Voyages*, p. 29.
36. C. Todd, 'French advertising in the eighteenth century', *Studies on Voltaire and the Eighteenth Century*, 266 (1989), pp. 513–4, 532, 544–45.
37. N. Coquery, 'French Court Society and Advertising Art: The Reputation of Parisian Merchants at the End of the Eighteenth Century', in C. Wischermann and E. Shore (eds), *Advertising and the European City: Historical Perspectives* (Aldershot: Ashgate, 2000), p. 97; and C. Walsh, 'The Advertising and Marketing of Consumer Goods in Eighteenth-Century London', in Wischermann and Shore (eds), *Advertising and the European City: Historical Perspectives* (Aldershot: Ashgate, 2000), pp. 79–95, p. 84. On the ways in which advertising targeted the middling sorts see C. Jones, 'The Great Chain of Buying: Medical Advertisement, the Bourgeois Public Sphere, and the Origins of the French Revolution', *American Historical Review*, 101 (1996), pp. 13–40. On advertising more generally see G. Feyel, 'Négoce et presse provinciale en France au 18e siècle: methods et perspectives de recherches' in F. Angiolini and D. Roche (eds), *Cultures et formations négociantes dans l'Europe moderne* (Paris: Editions de l'Ecole des Hautes Etudes en Sciences Sociales, 1995), pp. 439–511. C. Wischermann, 'Placing Advertising in the Modern Cultural History of the City', in Wischermann and Shore (eds), *Advertising and the European City*, pp. 1–2.
38. J. Styles, 'Product Innovation in Early Modern London', *Past and Present*, 168 (2000), p. 132.
39. Ibid., p. 165.
40. Faujas de Saint-Fond (ed.), *Description des expériences aérostatiques*, vol. 1, pp. 7–9.
41. Ibid., p. 35.
42. Rivarol, "Lettre à Monsieur le Président de ***'.

43. *Courrier d'Avignon*, 9 September 1783, p. 288.
44. *Correspondance littéraire, philosophique et critique*, 13 (August 1783), pp. 347–8.
45. *Journal de Paris*, 29 April 1784, p. 525; and ibid., 2 October 1784, pp. 1167.
46. Ibid., 26 November 1783, p. 1356 and ibid., 28 November 1783, pp. 1363–4.
47. *Journal de Nancy*, 12:17 (1783), pp. 47–8.
48. *Morning Herald and Daily Advertiser*, 24 November 1783; *Gentleman's Magazine*, November 1783, p. 977.
49. *Bristol Journal*, 3 January 1784, p. 3; *Morning Herald and Daily Advertiser*, 18 December 1783; ibid., 24 November 1783; *Gazetteer and New Daily Advertiser*, 6 December 1783; ibid., 19 December 1783; and *Morning Post*, 18 December 1783.
50. *Morning Herald and Daily Advertiser*, 16 December 1783.
51. *Affiches d'Angers*, 27 August 1784, p. 147. For the list of subscribers see *Journal Encyclopédique*, November 1784, pp. 85–6. [P. Lévêque], *Déscription de la second expérience aérostatique faite à Nantes, le 6 Septembre 1784* (Nantes: De Brun, [1784]).
52. *Affiches de Province*, 20 March 1784, pp. 171–2. Also see Gattey, Megnié, and Henry, *Souscription pour la construction d'un vaisseau aérien* (N.p., n.d.).
53. *Journal de Paris*, 4 April 1784, p. 419.
54. *Courrier d'Avignon*, 9 April 1784, p. 120.
55. *Affiches de Toulouse et du Haut-Languedoc*, 4 February 1784, p. 21.
56. *Affiches de Paris*, 7 March 1784, 622; *Journal de Paris*, 29 April 1784, p. 525.
57. The same advertisement appeared in the *Journal de Paris*, 2 October 1784, p. 1167 and the *Affiches de Paris*, 11 October 1784, p. 2679.
58. *Bath Chronicle*, 15 January 1784, p. 3.
59. Advertisement from the *Edinburgh Evening Courant* cited in Fergusson, *Balloon Tytler*, p. 63.
60. *Bristol Journal*, 3 January 1784, p. 3.
61. *Evening Post*, 7 September 1813, [p. 3].
62. On the exotic see, for example, D. Outram, *The Enlightenment* (Cambridge: Cambridge University Press, 1995), esp. ch. 5; and G. S. Rousseau and R. Porter, *Exoticism and the Enlightenment* (Manchester: Manchester University Press, 1990).
63. *Gazetteer and New Daily Advertiser*, 30 April 1784 (pyramid); ibid., 11 May 1784, p. 1 (Chinese inspired).
64. [Francesco Zambeccari], *Proposal for Making a Large Aerostatic Globe*, 1784.
65. *Gazetteer and New Daily Advertiser*, 17 February 1784; and *Morning Herald and Daily Advertiser*, 23 February 1784.
66. *Morning Post*, 4 August 1784.
67. Rolt, *The Aeronauts*, p. 65.
68. *Morning Post*, 16 August 1784; *Gazetteer and New Daily Advertiser*, 17 August 1784; *Morning Herald and Daily Advertiser*, 14 August 1784.
69. Lunardi, *An Account of the First Aerial Voyage*, p. 10.
70. *Morning Post*, 17 September 1784. On the possible confusion about locations see Ibid., 8 September 1784.
71. Ibid., 16 September 1784 (italics in original).
72. *Bath Chronicle*, 30 September 1784, p. 1.
73. *Morning Post*, 20 September 1784; *Gazetteer and New Daily Advertiser*, 21 September 1784; and *Morning Herald and Daily Advertiser*, 22 September 1784.
74. B. Sheridan, *Betsy Sheridan's Journal* (Oxford: Oxford University Press, 1986), p. 24.

75. *Morning Herald and Daily Advertiser*, 15 November 1784; ibid., 18 November 1784; *Bath Chronicle*, 7 October 1784, p. 2. The *Bath Chronicle* claimed that 2,500 people had visited the Pantheon on 1 October 1784. Sheridan, *Betsy Sheridan's Journal* p. 24. The price in the periodical press for this event was set at five shillings while Sheridan suggests it was just one shilling; her passage, however, may have been proverbial in nature. The *Morning Post* compared this success to that of Blanchard who was showing a balloon in London at the same time. Apparently, Blanchard's exhibitor had run off with the receipts amounting to as much as eighty pounds sterling. *Morning Post*, 2 October 1784.
76. *American Citizen*, 20 June 1806, [p. 2].
77. L. P. de Ségur, *Memoirs and Recollections* (London: H. Colburn, 1825), p. 31.
78. *Aerostation out at Elbows, or the Itinerant Aeronaut* [1785].
79. *Affiches de Paris*, 18 October 1785, pp. 2790–1; on Enslen see Oettermann, 'Die Fliegenden Plastiken des Johann Karl Enslen'.
80. *Courier de l'Europe*, 4 July 1786, p. 8.
81. On his later appearances in Paris see, for example, *Journal de Paris*, 24 September 1786, 1105. On his launches in other cities see *Affiches de Trois-Evêches*, 27 December 1787, p. 410 (for Metz); and *L'Année littéraire* (1785), p. 142 (for Strasbourg).
82. *Argus*, 27 June 1796, [p. 3].
83. Alban and Vallet, *Prècis des experiences faites par MM. Alban & Vallet; & Souscription propose pour un Cours de Direction Aérostatiques* (Paris: Velade; Javel: Chartier, 1785).
84. *Freeman's Journal and Daily Advertiser*, 30 December 1784–1 January 1785.
85. *Morning Post*, 9 May 1785.
86. Blanchard and Baker, *The Principles, History and Use of Air-Balloons*, p. 37.
87. *Morning Herald and Daily Advertiser*, 21 May 1785, n.p.
88. *Commercial Advertiser*, 30 September 1817, [p. 3].
89. *Baltimore Patriot*, 11 May 1818, [p. 3].
90. *Affiches de Trois-Evêches*, 25 January 1787, 26.
91. *Pennsylvania Packet*, 17 June 1784, [p. 3].
92. *Providence Gazette*, 1 October 1803, [p. 3].
93. For the launch without price see *Alexandria Gazette*, 2 September 1818, [p. 3]; for one of his usual prices see *National Messenger*, 11 September 1818, [p. 3].
94. *Freeman's Journal and Daily Advertiser*, 23–25 June 1785.
95. On the sheep see Hodgson, *The History of Aeronautics in Great Britain*, p. 172.
96. On his companions see Blanchard, *Liste chronologique*. This only goes up to 1803. For the parachuting dogs see Coutil, *Jean-Pierre Blanchard*, pp. 13–16. On Garnerin, see Poniatowski, *Garnerin*. On Garnerin and his dog see *Chronique de Paris*, 24 June 1790, 700.
97. *Morning Herald and Daily Advertiser*, 2 September 1785.
98. *City Gazette*, 25 January 1820, [p. 2]. Another aeronaut also offered half price tickets for children. See *Newport Mercury*, 31 August 1805, [p. 3].
99. *Massachusetts Mercury*, 19 December 1800, [p. 3].
100. *Affiches de Province*, 24 June 1786, 298–9 (endurance); *Reimpresson de l'Ancien Moniteur*, 18 October 1798 and ibid, 24 October 1798 (horse). For Tetu du Brissy see *Le cirque olympique, ou les exercices des Chevaux de MM. Franconi, du Cerf Coco, du Cerf Azor, de l'Eléphant Baba, suivi du Cheval Aéronaute, de M. Testu Brissy* (Paris: Nepveu, 1817); and M. Tetu du Brissy, *Sur l'exposition de l'Expérience aérostatique faite le 18 juin 1786* (Paris, 1786).

101. For Enslen and Ruggieri, see *Affiches de Paris*, 18 October 1785, pp. 2790–1; and *Journal de Paris*, 23 October 1785, 1219–20. On the balloons launched for charity see *Journal de Paris*, 28 March 1789, 396; and ibid., 11 April 1789, 458. On fireworks in France more generally see Lynn, 'Sparks for Sale'.
102. *Providence Gazette*, 4 September 1790, [p. 3].
103. Poterlet, *Notice sur Madame Blanchard.*
104. On Elisa Garnerin see Caron and Gével, 'Mademoiselle Elisa Garnerin'. On the Sadlers see Canby, 'James Sadler, Aeronaut'. On Robertson see E.-G. Robertson, *Mémoires récreatifs, scientifiques et anexdotiques*, 2 vols (Paris: Chez l'auteur, 1831–3).
105. On the accusation of greed see *Affiches de Province*, 30 October 1784, p. 615.
106. On his somewhat grandiose claims see, e.g., *Morning Herald and Daily Advertiser*, 8 November 1784, where Blanchard writes that he invented the parachute as early as 1777 a claim disputed by, among others, Thomas Martyn; see ibid., 10 November 1784.
107. On Marie-Madeleine-Sophie Blanchard see Poterlet, *Notice sur Madame Blanchard*; Coutil, *Jean-Pierre Blanchard*, pp. 20–4; Schneider, 'Star Balloonist of Europe', pp. 697–711; and Ferrero, 'Sophie Blanchard, Amazzone del Cielo, in Val Trebbia'.
108. Blanchard and Baker, *The Principles, History and Use of Air-Balloons*, p. 45.
109. Ibid., p. 43.
110. Jeffries, *A Narrative of Two Aerial Voyages*, pp. 39–41.
111. *Gazetteer and New Daily Advertiser*, 29 June 1785.
112. See, for example, Blanchard, *Journal and Certificates on the Fourth Voyage of Mr. Blanchard*; idem., *Journal of My Forty-Fifth Ascension*; and idem., *Relation du seizieme voyage aërien de M. Blanchard, fait à Gand, le 20 Novembre 1786* (Gand, 1786).
113. Rolt, *The Aeronauts*, p. 85.
114. For France see, e.g., Schwartz, *Spectacular Realities*; Michael B. Miller, *The Bon Marché: Bourgeois Culture and the Department Store, 1869–1920* (Princeton, NJ: Princeton University Press, 1981); and R. H. Williams, *Dream Worlds: Mass Consumption in Late Nineteenth-Century France* (Berkeley, CA: University of California Press, 1982).
115. On balloons as sublime, a pun combining the aesthetic nature of launch with the chemical process of sublimation, see Sgard, 'Les philosophes en montgolfière'.

Chapter 6

1. de Vries, *The Industrious Revolution.*
2. J. Brewer and F. Trentmann, 'Introduction', in *Consuming Cultures, Global Perspectives: Historical Trajectories, Transnational Exchanges* (Oxford: Berg, 2006), p. 3.
3. *Freeman's Journal*, 5–7 May 1785, n.p.
4. *Massachusetts Centinel*, 14 July 1784, [p. 3].
5. *Affiches de Province*, 20 Aprils 1786, pp. 187–8.
6. Rolt, *The Aeronauts*, p. 54.
7. On Guyot see G. Chabaud, 'Entre sciences et sociabilités: les experiences de l'illusion artificielle en France à la fin du XVIIIe siècle', *Bulletin de la Société d'Histoire Moderne et Contemporaine*, 44 (1997), pp. 36–44; and idem., 'La physique amusante et les jeux expérimantaux en France au XVIIIe siècle', *Ludica*, 2 (1996), pp. 61–73.
8. M. Guyot, *Essai sur la construction des ballons aérostatique et sur la maniere de les diriger* (Paris: Gueffier, 1784), n.p.
9. *Courrier d'Avignon*, 23 September 1783, p. 303.
10. *Journal de Paris*, 5 November 1783, p. 1261.

11. J.-C. Pingeron, *L'art de faire soi-même les ballons aérostatiques, conformes à ceux de M. de Montgolfier* (Paris: Hardouin, 1783), pp. 20–1.
12. *Almanach des Ballons, ou Globes Aérostatiques; Etrennes du jour physico-historique et chantantes* (Paris: Langlois, 1784), p. 76.
13. For Blondy's early advertisement see *Journal de Paris*, 17 September 1783, p. 1073.
14. *Affiches de Paris*, 8 May 1791, p. 1728. It is curious that as late as 1791 Lebrun would reveal his earlier connection with the royal family.
15. P.-J.-B. Nougaret, *Tableau mouvant de Paris, ou Variétés amusantes*, 3 vols (Paris: Duchesne, 1787), vol. 1, p. 153.
16. *Morning Herald and Daily Advertiser*, 6 September 1784, n.p.
17. Ibid., 2 January 1784, n.p.
18. Ibid., 10 February 1784, n.p.; also see Ibid., 12 January 1784, n.p.; Ibid., p. 6 February 1784, n.p.; and Ibid., 15 November 1784, n.p.
19. *Morning Chronicle*, 20 September 1784, n.p.
20. *Gazetteer and New Daily Advertiser*, 26 August 1784, n.p. Also see *Morning Post*, 28 August 1784, n.p.; *Morning Herald and Daily Advertiser*, p. 1 September 1784, n.p.; and *Publick Ledger*, 7 September 1784, n.p.
21. *Morning Herald and Daily Advertiser*, 22 September 1784, n.p.
22. *Gazetteer and New Daily Advertiser*, 16 November 1784, n.p.
23. Faure, *Les frères Montgolfier*, p. 192.
24. Hunn, 'The Balloon Craze in France', pp. 151–2.
25. Arecco, *Mongolfiere, science et lumi*, p. 82.
26. *Morning Post*, 30 October 1784, n.p.
27. For a description of a wide array of balloon-related objects see *Le Temps des Ballons: Art et Histoire* (Paris: Editions de La Martinière, 1994), pp. 42–84. This work contains many pictures, most of which are of objects on display at the Musée de l'Air et d'Espace in Le Borget, France.
28. *Courrier d'Avignon*, 21 October 1783, p. 336.
29. *American Herald*, 28 June 1784, [p. 1]; ibid., 7 June 1784, [p. 3].
30. *Spectator*, 4 September 1802, [p. 3].
31. *Norwich Packet*, 28 July 1785, [p. 3].
32. *Morning Herald and Daily Advertiser*, 7 October 1784, n.p.; and *Publick Ledger*, 7 October 1784, n.p.
33. Rolt, *The Aeronauts*, p. 72.
34. *New Spectator*, 9 March 1784, pp. 3–4.
35. *Morning Herald and Daily Advertiser*, 17 March 1784, n.p.
36. *Bath Chronicle*, 17 June 1784, p. 2.
37. *Morning Post*, 4 October 1784, n.p.
38. *New York Packet*, 3 October 1785, [p. 4].
39. Ibid., 3 October 1785, [p. 4].
40. Faure, *Les frères Montgolfier*, p. 189.
41. *Affiches de Toulouse et du Haut-Languedoc*, 4 February 1784, pp. 20–1.
42. *Morning Herald and Daily Advertiser*, 19 February 1784, n.p.
43. Rivarol, "Lettre à Monsieur le Président de ***'. Criticizing fashion was not, of course, new at the time of ballooning. See, for example, C. de Peyssonnel, *Petite Chronique du Ridicule*, ed. Mario Pasa (1782; Paris: Payot, 2007), pp. 131–5.
44. *Gazetteer and New Daily Advertiser*, 12 February 1784, n.p.
45. *Massachusetts Mercury*, 29 July 1800, [p. 3].

46. *Morning Post*, 12 October 1784, n.p.; also *Morning Herald and Daily Advertiser*, 20 October 1784, n.p.
47. *Morning Herald and Daily Advertiser*, 4 December 1784, n.p.
48. *Publick Ledger*, 27 August 1784, n.p.
49. Faure, *Les frères Montgolfier*, p. 189.
50. Ibid., p. 192. A similar game also appeared in New York. See *New York Packet*, 17 February 1785, n.p.
51. P. Guth, 'Tous les arts se sont mis "au ballon"', *Connaissance des Arts*, 59 (1957), p. 15.
52. *New York Packet*, 17 February 1785, n.p. Balloon gingerbread was also available in London; see *Publick Ledger*, 27 August 1784, n.p.
53. Hunn, 'The Balloon Craze in France', p. 151.
54. *Morning Post*, 19 April 1786, n.p.
55. Arecco, *Mongolfiere, science et lumi*, pp. 81–2; and Hunn, 'The Balloon Craze in France', p. 151.
56. *Columbian Herald*, 4 January 1785, [p. 1].
57. See some of the images in Guth, 'Tous les arts se sont mis "au ballon"', pp. 18–19.
58. Faure, *Les frères Montgolfier*, pp. 188–9.
59. Rolt, *The Aeronauts*, pp. 22–3.
60. *Columbian Centinel*, 4 July 1785, [p. 2].
61. *Commercial Advertiser*, 25 May 1808, [p. 3].
62. See some of the images in Guth, 'Tous les arts se sont mis "au ballon"', pp. 16–17.
63. Faure, *Les frères Montgolfier*, p. 189. On porcelain see also, Arecco, *Mongolfiere, science et lumi*, pp. 82–3. The Musée de l'Air et de l'Espace (Le Borget, France) has a balloon chamber pot on exhibit.
64. See the images in Guth, 'Tous les arts se sont mis "au ballon"', p. 18. Some of these tiles were reproduced in the late twentieth century and sold at the Metropolitan Museum of Modern Art in New York City, New York.
65. H. Rosenau, 'The Sphere as an Element in the Montgolfier Monuments', *Art Bulletin*, 50 (1968), pp. 65–6 and plates.
66. For a list of some song titles see Tissandier, *Bibliographie aeronautique*, pp. 51–3. For the opera see the libretto by M. Blumhofer, *Die Luftschiffer, oder der Strafplanet der Erde* (Leipzig: J.A. Imhoff, 1787).
67. 'Chanson du ballon de M. de Montgolfier', Gimbel Collection, XP XL-2. This song is written in the local patois.
68. E. Leroudier, *Noëls et chansons en patois Lyonnais du XVIII siècle* (Lyon: Cumin and Mason, 1918), pp. 81–3.
69. 'The favourite Balloon Song, sung by Mr. Arrowsmith, at Vauxhall Gardens; the words by Mr. Pilon, set to Musick by Mr. Arne', in *Massachusetts Spy*, 13 October 1785, [p. 4].
70. *Pennsylvania Packet*, 11 October 1784, [p. 3].
71. J. Duhem, *Musée aéronautique avant Montgolfier* (Paris: Sorlot, 1943), p. 219, for other songs see pp. 191, 215–22.
72. Boucault, *Ils font ce qu'ils peuvent* (N.p., n.d.), p. 3.
73. *L'Aérostad [sic] de l'Abbé Miollan* (N.p., n.d.), p. 2.
74. *New York Daily Advertiser*, 6 February 1818, [p. 3].
75. Rolt, *The Aeronauts*, p. 24.
76. See, for example, R. Chartier, *Cultural History: Between Practices and Representations*, trans. L. G. Cochrane (Ithaca, NY: Cornell University Press, 1988); idem., *The Cultural Uses of Print in Early Modern France*; Darnton, *The Business of Enlightenment*; idem., *The*

Literary Underground of the Old Regime (Cambridge, MA: Harvard University Press, 1982); and idem., *The Forbidden Best-Sellers of Pre-Revolutionary France* (New York: Norton, 1996).

77. See, for example, on Newton and literature, M. H. Nicolson, *Science and Imagination* (Ithaca, NY: Great Seal Books, 1956). On balloons in literature see, for example, Heggen, 'Die "ars volandi"'; A. A.W. Coley, 'Followers of Daedalus: Science and Other Influences in the Tales of Flight in Eighteenth-Century French Literature', *Studies on Voltaire and the Eighteenth Century*, 371 (1999), pp. 81–173; and idem., 'The Theme of Experiment in French Literature of the Earlier Eighteenth Century (1715–1761)', *Studies on Voltaire and the Eighteenth Century*, 332 (1995), pp. 213–333.
78. Gay, *The Enlightenment: An Interpretatio*, vol. 2, p. 133.
79. For some Latin poetry see, for example, 'Carolo Greenio' in Mason, *Aeronautica*, p. 351; and the poem written after the death of Pilâtre de Rozier and Romains, 'Pro diro & infelicissimo funere PILATRII ROZERII', in *Journal encyclopédique*, December 1785, p. 445.
80. Tourneux (ed.), *Correspondance littéraire, philosophique et critique*, vol. 14, p. 165.
81. P. Shelley, *The Esdaile Notebooks: A Volume of Early Poems*, ed. K. N. Cameron (New York: Knopf, 1964), p. 89.
82. For a nice analysis of the poem see E. Kraft, 'Anna Letitia Barbauld's "Washing Day" and the Montgolfier Balloon', *Literature and History*, 4:2 (1995), pp. 25–41. On Barbauld more generally see W. McCarthy, *Anna Letitia Barbauld: Voice of the Enlightenment* (Baltimore, MD: The Johns Hopkins University Press, 2008).
83. E. Darwin, *The Botanic Garden*, ed. D. H. Reiman, 2 vols (New York: Garland, 1978), vol. 1, p. 174.
84. *Aris's Birmingham Gazette*, 10 January 1785, n.p.
85. Poem cited in B. T. Brown, 'French Scientific Innovation in Late- Eighteenth-Century Dublin: The Hydrogen Balloon Experiments of Richard Crosbie (1783-1785)', in Gargett and Sheridan (eds), *Ireland and the French Enlightenment, 1700–1800* (New York: St. Martin's, 1999), pp. 107–26, p. 118.
86. *Almanach des muses* (1784), pp. 25–7.
87. Ibid., p. 194. This was reprinted in the *Journal encyclopédique* 57 (1784), vol. 1, p. 293.
88. *Almanach des muses* (1784), p. 52. Poems appeared across Europe honouring the Montgolfiers. See, for example, 'Schiffer Bergbuchter, alias Montgolfier', in *Deutsches Museum*, March 1784, pp. 275–9.
89. *Lines on the Lamented Death of Mr. Sadler*, Gimbel Collection, XC-10-3B (Sad) – 3D.
90. *Morning Post*, 18 November 1784, n.p.
91. 'Sul volo del Capitan Lunardi sul Globo aerostatico esequito in Palermo nel di 31 Luglio 1790', in *Miscellnea di opuscoli su di varie materie, che nella sequente*, Newberry Library, Vault Ayer MS 1167, fol. 619.
92. Ballad cited in Rolt, *The Aeronauts*, p. 60.
93. *Morning Post*, 1 October 1784, n.p.
94. Ibid., 21 September 1784, n.p.
95. De Barbosa, du Bocage, *Elogio Poetico*, p. 7.
96. J. Tytler, 'To Mr. Lunardi, of the Successful Aerial Voyages', in Vincenzo Lunardi, *An Account of Five Aerial Voyages in Scotland* (London, 1786), pp. 105–15, cite on 105.
97. *Morning Post*, 19 January 1785, n.p.
98. J.-P. Blanchard, 'Stances', Duchosal, *Blanchard, poème en quatre chants.*
99. Duchosal, *Blanchard, poème en deux*; idem., *Blanchard, poème en quatre chants.*

100. *Affiches de Province*, 8 January1784, p. 15.
101. *Morning Herald and Daily Advertiser*, 1 October 1784, n.p.
102. Arnaud de Saint-Maurice, *L'Observatoire volant et le triomphe héroïque de la navigation aérienne, et des vésicatoires amusants et célestes, poëme en quatre chants* (Paris: Cussac, 1784), pp. 19–20.
103. V. Lancetti, *Areostiade: ossia, il mongolfiero, poema*, 2 vols (Milan: Agnello Nobile, 1803). See also the somewhat shorter *Le siècle des ballons, satyre nouvelle* (Paris: Cailleau, 1784).
104. *Pennsylvania Evening Herald*, 15 March 1786, p. 57 (Seddon); *Pennsylvania Packet*, 31 May 1785, [p. 3] (Woodhouse). Woodhouse was still advertising this poem in September 1785. See Ibid., 7 September 1785, [p. 1].
105. *The Gazetteer and New Daily Advetiser*, 20 October 1783, n.p.
106. *Daily Advertiser*, 9 April 1803, [p. 2].
107. *Morning Chronicle*, 21 May 1803, [p. 2].
108. *Independent Gazetteer*, 17 April 1784, [p. 3]. Reprinted in the *Weekly Museum*, 10 October 1789, [p. 4].
109. *Weekly Wanderer*, 17 November 1806, [p. 4].
110. *Morning Herald and Daily Advertiser*, 2 September 1784, n.p.
111. Printed in Hodgson, *The History of Aeronautics in Great Britain*, p. 140.
112. *Morning Post*, 18 October 1784, n.p
113. Ibid., 16 October 1784, n.p.
114. P. Keen, The "Balloonomania": Science and Spectacle in 1780s England', *Eighteenth-Century Studies*, 39 (2006), pp. 507–35 pp. 530–1.
115. E. Inchbald, *The Mogul Tale, or The Descent of the Balloon* (London: Powell, 1796), p. 4. On British balloon literature of all kinds see Keen, 'The "Balloonomania"'.
116. Inchbald, *The Mogul Tale*, p. 5.
117. Ibid., p. 6.
118. Ibid., p. 21.
119. *New York Gazette*, 9 April 1798, [p. 3].; *Boston Price-Current*, 31 July 1797, n.p.
120. F. Pilon, *Aerostation; or, The Templar's Stratagem* (London: Kearsley, 1784).
121. Ibid., p. 32.
122. *London Chronicle*, 28–30 October 1784, p. 403. Cf. the brief review in *Monthly Review*, December 1784, p. 476.
123. These two plays could be the same, but with different titles. See *Morning Post*, 17 July 1784, n.p.; *Morning Post*, 16 August 1784, n.p.; and *Morning Herald and Daily Advertiser*, 30 June 1785, n.p.
124. *Morning Herald and Daily Advertiser*, 19 December 1783, n.p.
125. N.-M.-F. Bodard de Tezay, *Le ballon, ou La physicomanie* (Paris: Galande, 1783).
126. N.-F.-M. Bodard de Tezay, *Arlequin, roi dans la lune* (Paris: Cailleau, 1784).
127. See K. Turner, 'The Spectacle of Democracy in the Balloon Plays of the Revolutionary Period', *Forum for Modern Language Studies*, 39 (2003), pp. 241–53. L.-A. Beffroy de Reigny, *Nicodème dans la lune, ou la revolution pacifique*, 3rd edn (Paris: Froullé, 1791).
128. D. Boutard, *La femme en parachute, ou Le Soupçon* (Paris: Chez la Librairie au Théâtre du Vaudeville, 1800).
129. J.-B.-A. Hapdé and H. Chaussier, *Le Parachute, comédie-parade en un acte et en prose* (Paris: Sages, An VI).
130. *Journal du Hainaut*, 1788, pp. 8–9.

131. *The Balloon or Aerostatic Spy, A Novel, Containing a Series of Adventures of an Aerial Traveller; Including a Variety of Histories and Characters in Real Life*, 2 vols (London: W. Lane, 1786).
132. Ibid., vi.
133. D. Moore, *An Account of Count d'Artois and his Friend's Passage to the Moon, In a Flying Machine, called, an Air Balloon* (Litchfield, CT: Collier, 1785). A version of this book was also published in America; see the announcement in *Vermont Gazette*, 5 June 1786, [p. 3].
134. *Aventures singulères, d'un voyageur aérien* (London: Hookham; Paris: Duchesne, 1785).
135. J.-P. Richter, *Journal de bord de l'aéronaute Giannozzo/Des Luftschiffers Giannozzo Seebuch*, ed. and trans. Jacques Poumet (Paris: Aubier Montaigne, 1975).
136. For *L'Amour dans le Globe* see the *Affiches de Province*, 25 January 1784, p. 109; *Le Mouton, le Canard, et le Coq. Fable dialoguée* (Bruxelles, 1783); *Réclamation du Mouton, premier navigateur aërien* (Paris, Cailleau, 1784); and *Histoire intéressante d'un nouveau voyage à la lune, et de la descente à Paris d'une jolie dame de cette terre étrangère* (Paris: F.-G. Deschamps, 1784).
137. F. Suleau, *Voyage en l'air: second réveil* ([Paris, 1791]).
138. [C. Bretin], *Contes en vers* (Paris: Gueffier, 1799).
139. *Alida et Dorval, ou la Nymph de l'Amstel* (Veropolis, 1785).
140. *Le Retour de mon pauvre oncle, ou relation de son voyage dans la lune* (Paris: Lejay, 1784). Cf. Voltaire, 'Micromegas', in *Candide and Other Stories*, trans. Roger Pearson (Oxford: Oxford University Press, 1990), pp. 101–21; and Cyrano de Bergerac, *Voyages to the Moon and the Sun*.
141. *Journal de Paris*, 3 October 1783, 1141; see also *Massachusetts Spy*, 15 January 1784, p. 1.
142. *Bangor Weekly Register*, 20 January 1820, [p. 4].
143. *Morning Chronicle*, 24 March 1803, [p. 2].
144. *New Jersey Journal*, 3 June 1789, [p. 3].
145. *Independent Ledger*, 17 January 1785, [p. 3].
146. *Giornale aerostatico* (1784).
147. The official section title in the *Journal encyclopédique* was "Notices diverses concernant la machine aerostatique, &c, extraits des papiers publics."
148. *The Balloon Almanac* (1787), advertised in *Freeman's Journal*, 3 January 1787, [p. 4] including a list of the table of contents. Also see Ibid., 14 September 1785, [p. 3]; and Ibid., 28 January 1789, [p. 4].
149. J. F. Pilâtre de Rozier, *Première expérience de la Montgolfiére, construite par ordre du roi* (Paris: L'Imprimierie de Monsieur, 1784); and Robert and Robert, *Mémoire sur les Expériences Aérostatiques*; and F. Laurent d'Arlandes and J.-A.-C. Charles, *Les voyageurs aëriens* (N.p., 1784).
150. See, for example, J.-P. Blanchard, *Abrégé de mes aventures terrestres depuis le mois de janvier 1787* (N.p., n.d.); idem., *An Exact and Authentic Narrative of M. Blanchard's Third Aerial Voyage* (London: Heydinger, 1784); idem., *Journal and Certificates on the Fourth Voyage of Mr. Blanchard* (London: Baker and Galabin, 1784); idem., *Journal of My Forty-Fifth Ascension*; idem., *Procès-verbaux lors du troisieme voyage aërien de M. Blanchard, accompagné de B. Boby, gressier au Parlement, fait le 18 Juillet 1784* (n.p., n.d.); idem., *Relation du quinzieme voyage aërien de M. Blanchard*; and idem., *Relation du seizieme voyage aërien de M. Blanchard, fait à Gand, le 20 Novembre 1786.*
151. Blanchard and Baker, *The Principles, History and Use of Air-Balloons.*

152. Lunardi, *An Account of the First Aerial Voyage in England*; and idem., *An Account of Five Aerial Voyages in Scotland* (London, 1786).
153. *Morning Post*, 25 September 1784, n.p.
154. Baldwin, *Airopaidia.*
155. Cavallo, *The History and Practice of Aerostation.*
156. Faujas de Saint-Fond (ed.), *Description des expériences aérostatiques.*
157. Martyn, *Hints of Important Uses*; Guyot, *Essai sur la construction des ballons aérostatique et sur la maniere de les diriger*; Amoretti, *Delle machine aerostatich*; and Bertholon, *Des avantages que la physique.* Also see A. Deparcieux, *Dissertation sur les globes aérostatiques* (Paris: Chez l'Auteur, 1783).
158. M.-J. Brisson, *Dictionnaire de la physique*, 3 vols (Paris, 1790). For the American encyclopedia see *Pennsylvania Packet*, 3 March 1790, [p. 3].

Conclusion

1. E. A. Poe, 'The Balloon Hoax' in *Sixty-Seven Tales* (New York: Gramercy Books, 1985), pp. 418–25; J. Verne, *Five Weeks in a Balloon* (Leydon, MA: Aeonian Press, 1976).
2. R. Holmes, *The Age of Wonder: How the Romantic Generation Discovered the Beauty and Terror of Science* (New York: Pantheon, 2008), pp. xvii–xviii. More generally see Holmes, ch. 3.
3. On Romanticism and science see A. Cunningham and N. Jardine (eds) *Romanticism and the Sciences* (Cambridge: Cambridge University Press, 1990).
4. For the use of balloons in arctic exploration see M. Robinson, *The Coldest Crucible: Arctic Exploration and American Culture* (Chicago, IL: University of Chicago Press, 2006), pp. 114–15.
5. M. Thébaud-Sorger, 'Innovation and Risk Management in Late Eighteenth-Century France: The Administration of Inventions in French Cities at the End of the Old Regime', in C. Rabier (ed.), *Fields of Expertise: A Comparative History of Expert Procedures in Paris and London, 1600 to Present* (Newcastle: Cambridge Scholars Publishing, 2007), pp. 268–72.
6. See, Lynn, *Popular Science and Public Opinion in Eighteenth-Century France*, ch. 5.
7. Veblen, *The Theory of the Leisure Class* (1899; New York: Penguin, 1994); G. Simmel, 'Fashion', *The International Quarterly*, 10 (October 1904), pp. 130–55; Elias, *The Court Society*, trans. E. Jephcott (New York: Pantheon, 1983); McKendrick, Brewer, and Plumb, *The Birth of a Consumer Society.*
8. On Lunardi and the Prince of Wales see, e.g., *Morning Post*, 16 September 1784.
9. See, for example, Potain's published list of subscribers in *Freeman's Journal*, 4–7 June 1785, which included the Duke of Leinster, the Earl of Moira, eight lords along with a number of other people.
10. M. Kwass, 'Big Hair: A Wig History of Consumption in Eighteenth-Century France', *American Historical Review*, 111 (2006), pp. 631–59. For a critique of the emulation thesis see P. Glennie, 'Consumption within Historical Studies', in D. Miller (ed), *Acknowledging Consumption: A Review of New Studies*, (London: Routledge, 1995), p. 179; and S. Nedadic, 'Middle-Rank Consumers and Domestic Culture in Edinburgh and Glasgow, 1720–1840', *Past and Present*, 145 (1994), pp. 122–56, esp. 124.
11. See, for example, J. Baudrillard, *The Consumer Society: Myths and Structures* (London: Sage, 1998). More generally, see the essays in *The Social Life of Things: Commodities in Cultural Perspective*, ed. A. Appadurai (Cambridge: Cambridge University Press, 1986).

12. Appleby, 'Consumption in Early Modern Social Thought', in *Consumption and the World of Goods*, pp. 162–73.
13. Smith, *Consumption and the Making of Respectability, 1600–1800* (London: Routledge, 2002); also see J. Jones, *Sexing la Mode: Gender, Fashion and Commercial Culture in Old Regime France* (Oxford: Berg, 2004).
14. Campbell, *The Romantic Ethic and the Spirit of Modern Consumerism* (London: Blackwell, 1987). Cf. with historian Jessica Riskin who has argued for the 'sensible' nature of science; Riskin, *Science in the Age of Sensibility: The Sentimental Empiricists of the French Enlightenment* (Chicago, IL: University of Chicago Press, 2002).
15. *Supplément à l'art de voyager dans les airs, contenant le Précis historique de la grande Expérience faite à Lyon le 19 Janvier 1784 et l'Exposé d'un moyen ingénieux pour diriger a volonté les Ballons aérostatiques* (N.p., n.d), p. 17.
16. C. Campbell, 'Understanding Traditional and Modern Patterns of Consumption in Eighteenth-Century England: A Character-Action Approach', in *Consumption and the World of Goods*, pp. 40–57.
17. Zambeccari, for example, describes the 'pleasure' derived from watching balloons take off; see *Morning Post*, 19 March 1785.

WORKS CITED

Manuscript Sources

Archives de l'Académie des Sciences (Institut de France):

Dossiers on members

Bibliothèque de l'Institut de France:

Fonds Joseph Bertrand

Musée de l'Air et de l'Espace, Le Bourget:

Fonds Montgolfiers

Newberry Library, Chicago, IL (USA):

"Miscellanea di ouscoli su di varie materie, che nella sequente." Vault Ayer MS 1167.

United States Air Force Academy, Colorado Springs, CO (USA):

Gimbel Collection

Periodicals

Some periodicals were examined using the Early American Newspapers collection which includes over seven hundred newspapers in an electronic, searchable database.

Affiches d'Angers.

Affiches d'Artois, le Boulonnois et le Calaisis.

Affiches du Beauvaisis.

Affiches de Bordeaux.

Affiches de Dauphiné.

Affiches de l'Orléannais.

Affiches de la Basse-Normandie.

Affiches de la Province de Bretagne.

Affiches de Normandie.

Affiches de Paris.

Affiches de Poitou.

Affiches de province, ou Annonces, affiches et avis divers.

Affiches de Toulouse et du Haut-Languedoc.

Affiches des Trois-Evêches.

Affiches du Hainaut & du Cambresis. Also *Journal du Hainaut & du Cambresis.*

Albany Gazette.

Alexandria Advertiser.

Alexandria Gazette.

American Apollo.

American Citizen.

American Herald.

American Recorder.

Annales de Chimie.

L'Année littéraire.

Argus.

Aris's Birmingham Gazette.

Baltimore Patriot.

Bangor Weekly Register.

Bath Chronicle.

The Bee.

Berlinische Monatsschrift.

Bibliothèque physico-economique, instructive et amusant.

Boston Daily Advertiser.

Boston Gazette.

Boston Intelligencer.

Boston Patriot.

Boston Price-Current.

Bristol Journal.

Carlisle Gazette.

Carolina Federal Republican.

Centinel of Freedom.

Chester Chronicle, or Commercial Intelligencer.

Chronique de Paris.

City Gazette.

City of Washington Gazette.

Claypoole's American Daily Advertiser.

Columbian Centinel.

Columbian Herald.

Commercial Advertiser.

Connecticut Gazette.

Connecticut Herald.

Connecticut Journal.

Courier.

Courier de l'Europe.

Courier of New Hampshire.

Courrier d'Avignon.

Daily Advertiser.

Daily National Intelligencer.

Deutsches Museum.

Diary.

English Chronicle.

Essex Journal.

Essex Register.

European Magazine.

Evening Post.

Farmer's Cabinet.

Federal Gazette.

Freeman's Journal and Daily Advertiser.

Gazette of the United States.

Gazetteer and New Daily Advertiser.

General Evening Post.

Gentleman's Magazine.

General Advertiser.

Giornale aerostatico.

Glocester Journal.

Green Mountain Patriot.

Hive.

Independent Chronicle.

Independent Gazetteer.

Independent Ledger.

Jackson's Oxford Journal.

Journal de Nancy.

Journal de Paris.

Journal de physique.

Journal der Physik.

Journal des sçavans.

Journal du Licée de Londres, ed. J. P. Brissot de Warville.

Journal Encyclopédique.

Journal Gratuit.

Litchfield Monitor.

London Chronicle.

London Courant and Westminster Chronicle.

London Evening-Post.

London Packet, or, New Lloyd's Evening Post.

London Recorder, or Sunday Gazette.

Magazin für das Neueste aus der Physik und Naturgeschichte, 1781-1799.

The Mail.

Massachusetts Centinel.

Massachusetts Gazette.

Massachusetts Mercury.

Mercantile Advertiser.

Mercure de France.

Merrimack Magazine.

Middlesex Gazette.

Monthly Review.

Morning Chronicle.

Morning Herald and Daily Advertiser.

Morning Post.

National Advocate.

National Gazette.

National Messenger.

New England Gazette.

New Hampshire Gazette.

New Hampshire Patriot.

New Hampshire Spy.

New Haven Gazette.

New Jersey Gazette.

New Jersey Journal.

The New Spectator.

New York Commercial Advertiser.

New-York Daily Advertiser.

New-York Gazette.

New-York Packet.

Newburyport Herald.

Newport Mercury.

Minerva.

National Advocate.

National Gazette.

Norwich Packet.

Opuscoli scelti sulle scienze. 1778-1803.

Oracle of Dauphin.

Oracle of the Day.

Otsego Herald.

Pennsylvania Evening Herald.

Pennsylvania Packet.

Philadelphia Gazette.

Political Magazine.

Political Repository.

Portsmouth Oracle.

Post-Boy.

Poulson's American Daily Advertiser.

Providence Gazette.

Publick Advertiser.

Publick Ledger.

Register of the Times.

Republican Agriculturalist.

Reimpression de l'Ancien Montieur.

Scots Magazine.

Rhode Island American.

St. James's Chronicle, or, the British Evening Post.

Salem Gazette.

South Carolina Weekly Gazette.

Spectator.

State Gazette of South Carolina.

Time Piece.

Traveller.

United States Chronicle.

United States Gazette.

Virginia Journal.

Walker's Hibernian Magazine.

Weekly Messenger.

Weekly Museum.

Weekly Oracle.

Weekly Wanderer.

Windham Herald.

World.

Yankee.

Primary Printed Sources

L'Aérostad [sic] de l'Abbé Miollan (n.p., n.d).

Aerostation Displayed (London: Hammond, 1802).

The Air Ballon Chace: or Mr. Blanchard's Flying Vessal [1784].

The Air Balloon: Or a Treatise on the Aerostatic Globe, 4th edn (London: Kearsley, 1784).

The Air-Balloon; or, The Sages Adventures in a Flight to the Moon (London: G. Bremner, 1784).

Air Balloons: Aeronautica, or Voyages in the Air (London: Griffiths, 1802).

Alida et Dorval, ou la Nymph de l'Amstel (Veropolis, 1785).

Alban and Vallet, *Prècis des expériences faites par MM. ALBAN & VALLET; & Souscription proposée pour un Cours de Direction Aérostatiques* (Paris: Velade; Chartier; Javel, 1785).

Almanach des Ballons, ou Globes Aérostatiques; Etrennes du jour physico-historique et chantantes (Paris: Langlois, 1784).

Amoretti, C., *Delle machine aerostatiche*, ed. L. Pescasio (Mantova: Circolo Bibliofilo Aeronautico; Editoriale Padus, 1970).

L'amour physicien, ou l'origine des ballons (Paris: Cailleau, 1784).

Anglande, J., *Les Montgolfiers* (Paris: Perrin, 1990).

Anon. (ed.), *The Romance of Ballooning: The Story of the Early Aeronauts* (New York: Viking, 1971).

Arnaud de Saint-Maurice, *L'Observatoire volant et le triomphe héroïque de la navigation aérienne, et des vésicatoires amusants et célestes, poëme en quatre chants* (Paris: Cussac, 1784).

Les Animaux savants (Paris: Didot, 1816).

Aréostation. Establissement d'une compagnie aéronautique (n.p., 1790).

d'Arlandes, F. L. and J.-A.-C. Charles, *Les voyageurs aëriens* (n.p., 1784).

L'art de la guerre changé par l'usage des machines aérostatiques (n.p., 1784).

An Asylum for Fugitive Pieces in Prose and Verse, 4 vols (London: Debrett, 1785–93).

Aulard, F.-A. (ed.), *Paris pendant la reaction Thermidorean et sous le Directoire*, 5 vols (Paris, 1898–1902).

Aventures singulères, d'un voyageur aérien (London: Hookham; Paris: Duchesne, 1785).

Avertissement au Peuple, sur l'Enlevement des Ballons ou Globes en l'air (Paris: Herissant, 1783).

Avis: Expérience aérostatique (Paris: Cailleau, 1787).

Baldwin, T., *Airopaidia* (Chester: Printed for the Author, 1786).

The Balloon Jester; or, Flights of Wit and Humour (London: W. Lane, n.d).

The Balloon or Aerostatic Spy, A Novel, Containing a Series of Adventures of an Aerial Traveller; Including a Variety of Histories and Characters in Real Life, 2 vols (London: W. Lane, 1786).

Banks, J., *The Letters of Sir Joseph Banks: A Selection, 1768–1820*, ed. N. Chambers (London: Imperial College Press, 2000).

Barbauld, A. L., *The Works of Anna Laetitia Barbauld*. 2 vols (London: Longman, 1825).

Baring-Gould, W. S., and C. Baring-Gould, *The Annotated Mother Goose* (New York: New American Library, 1967).

Beaufoy, H. B H., *Journal kept by H.B.H.B. during an aerial voyage with Mr. James Sadler* (London: G. Woodfall, 1811).

Beffroy de Reigny, L.-A. *Nicodème dans la lune, ou la révolution pacifique*. 3rd edn (Paris: Froullé, 1791).

Behn, A., 'The Emperor of the Moon', in J. Spencer (ed.), *The Rover; The Feigned Courtesans; The Lucky Chance; The Emperor of the Moon* (Oxford: Clarendon, 1995), pp. 271–335.

Bertholon, P., *Des avantages que la physique, et les arts qui en dependent, peuvent retirer des globes aërostatiques* (Montpellier: Jean Martel, 1784).Bianchini, Francesco. *Don Paolo Andreani*. Bologna, 1784.

Blanchard, J.-P., *Abrégé de mes aventures terrestres depuis le mois de janvier 1787* (n.p., n.d.).

—, *L'aéronaute Blanchard aux dames de Valognes* (Valognes: Clamorgam, n.d.).

—, *An Exact and Authentic Narrative of M. Blanchard's Third Aerial Voyage* (London: Heydinger, 1784).

—, *Journal and Certificates on the Fourth Voyage of Mr. Blanchard* (London: Baker and Galabin, 1784).

—, *Relation du quinzieme voyage aërien de M. Blanchard, fait à Francfort sur le Meyn, le 3 Octobre 1785* (Francfort sur le Meyn, 1785).

—, *Relation du seizieme voyage aërien de M. Blanchard, fait à Gand, le 20 Novembre 1786* (Gand, 1786).

—, *Journal of My Forty-Fifth Ascension and the First in America*, ed. C. Frey (Philadelphia, PA: The Penn Mutual Life Insurance Company, 1943).

—, *Liste chronologique des ascensions de l'Aéronaute Blanchard* (n.p., n.d).

—, *Procès-verbaux lors du troisieme voyage aërien de M. Blanchard, accompagné de B. Boby, gressier au Parlement, fait le 18 Juillet 1784* (n.p., n.d).

Blanchard, J.-P., and G. Baker, *The Principles, History and Use of Air-Balloons* (New York: Van Alen, 1796).

Blumhofer, M., *Die Luftschiffer, oder der Strafplanet der Erde* (Leipzig: J.A. Imhoff, 1787).

Bocage, M. M. de B. du, *Elogio Poetico à admiravel intrepdiz, com que en Domingo 24 de Agosto de 1784* (Lisboa: Ferreira, 1794).

Bodard de Tezay, Nicolas-Marie-Félix. *Arlequin, roi dans la lune* (Paris: Cailleau, 1784).

—, *Le ballon, ou La physicomanie* (Paris: Galande, 1783).

Bourgeois, D., *Recherches sur l'art de voler, depuis la plus haute antiquité jusqu'à ce jour* (Paris: Cuchet, 1784).

Boutard, D., *La femme en parachute, ou Le Soupçon* (Paris: Chez la Librairie au Théâtre du Vaudeville, 1800).

[Bretin, C.], *Contes en vers* (Paris: Gueffier, 1799).

Brisson, Mathurin-Jacques. *Dictionnaire de la physique*. 3 vols (Paris, 1790).

—, *Observations sur les nouvelles découvertes aërostatiques et sur la probabilité de pouvoir diriger les Ballons* (Paris: Boucher & Lamy, 1784).

Brown, J.. *Select Remains of the Rev. John Brown* (Pittsburgh, PA: Cramer, Spear and Eichbaum, 1810).

Bruno, G. D., *L'Utile direzione delle machine aerostatiche* (Naples, 1784).

Carra, J.-L., *Essai sur la nautique aérienne* (Paris: Eugène Onfroy, 1784).

Castelli, C., *Relation des expériences de la machine aérostatique de Don Paul Andréani* (n.p., 1784).

Cavallo, T., *The History and Practice of Aerostation* (London, 1784).

Le cirque olympique, ou les exercices des Chevaux de MM. Franconi, du Cerf Coco, du Cerf Azor, de l'Eléphant Baba, suivi du Cheval Aéronaute, de M. Testu Brissy (Paris: Nepveu, 1817).

Costa, M. A., *Saggi sull'aerostatica e sull'aeronautica* (Naples: Fibreno, 1837).

Coutelle, J. M. J., *Sur l'aérostat employé aux armées de Sambre-et-Meuse et du Rhin* (Paris: Crapelet, 1829).

Cowper, W., *The Letters and Prose Writings of William Cowper*, 3 vols, ed. J. King and C. Ryscamp (Oxford: Oxford University Press, 1981).

Cradock, A. F., *Journal de Madame Craddock: Voyage en France (1783–1786)*, trans. and ed. O. Delphin Balleyguier (Paris: Perrin, 1896).

Cyrano de Bergerac, S., *Voyages to the Moon and the Sun*, trans. R. Aldington (New York: Orion, 1962).

Darwin, E., *The Botanic Garden*, ed. D. H. Reiman, 2 vols (New York: Garland, 1978).

Découverte d'un point d'appui dans l'air, à l'usage des machines aérostatiques pour naviguer contre le vent (Paris: Cailleau, 1784).

Defoe, D., *The Consolidator, or Memoirs of Sundry Transactions from the World in the Moon*, ed. M. J. Bosse (New York: Garland, 1972).

Degen, J., *Beschreibung einer neuen flugmaschine* (Vienna, 1808).

Delany, M., *The Autobiography and Correspondence of Mrs. Delany*, ed. S. Chauncey Woolsey (Boston, MA: Roberts Brothers, 1879).

Deparcieux, A., *Dissertation sur les globes aérostatiques* (Paris: Chez l'Auteur, 1783).

Descente en parachute par Mlle. Garnerin, Le 25 Octobre 1815 (n.p., n.d.).

Description de deux Machines propres à la Navigation aérienne (n.p., n.d.).

Description de la seconde expérience aérostatique faite à Nantes, le 6 septembre 1784 (Nantes: Brun, n.d).

A Dictionary of the Wonders of Art (London: J. Cundee, 1803).

Discorso sopra la navigazione aerea, 2nd edn (Venice: Domenico Fracasso, 1789).

Discours sur les découvertes en general, et particulièrement sur deux des principales découvertes de ce siècle (Paris: Philippe-Denys Pierres, 1784).

Dissertation sur les aerostats des anciens et des modernes (Geneva, 1784).

Duchosal, Marie-Emile-Guillaume. *Blanchard, poème en deux chants* (Rouen; Paris: Marchands de Nouveautés, 1784).

—, *Blanchard, poème en quatre chants* (Bruxelles, 1786).

Ehrmann, F.-L. and C. Würtz. *Montgolfier'sche Luftkörper oder aerostatische Maschinen, eine Abhandlung, worinn die Kunst sie zu verfertigen und die Geschichte der bisher damit angestellten Versuche beschrieben warden* (Strassburg: Treuttel, 1784).

Elisa Garnerin nel suo XXIII volo da Milano il XIX Aprile 1824 (Milan: Bettoni, 1824).

Estatta descrizione del globo e paracadute dell-aeronauta Madamigella Elisa Garnerin in occaione del suo volo esequito in Padova nell-anno 1825 (Padova: Camba, n.d.).

Fabry, D., *Réflexions sur la relation du voyage aérien de MM. Charles & Robert, et la brochure intitulée: Méthode aisée de faire la Machine aérostatique* (Paris: Chez les Libraries des Nouveautés, 1784).

Fars Fausse-Landry, Madame la Vicomtesse de, *Mémoires de Madame la Vicomtesse de Fars Fausse-Landry, ou Souvenirs d'une octogénaire*, 2 vols (Pairs: Ledoyen, 1830).

Faujas de Saint-Fond, B. (ed.) *Description des expériences aérostatiques de MM. De Montgolfier*, 2 vols (1784; Osnabrück: Otto Zeller, 1968).

—, *A Journey Through England and Scotland to the Hebrides in 1784*, ed. A. Geikie, 2 vols (Glasgow: Hugh Hopkins, 1907).

Fête aérostatique, qui sera célébrée aujourd'hui au Champs de Mars (n.p., n.d.).

Fête nationale, Qui sera célébrée aujourd'hui au Champ-de-Mars, aux Champs-Elysées, à la Halle, & sur la place de la Bastille. [Paris, *c.* 1790].

Franklin, B., *Benjamin Franklin on Balloons: A Letter Written from Passy, France, January Sixteenth MDCCLXXIV* (Saint Louis, MO: W.K. Bixby, 1922).

Freneau, P. M., *The Poems of Philip Freneau*, ed. F. L. Pattee (Princeton, NJ: The University Library, 1902–7).

Galien, J., *L'art de naviger dans les airs*. 2nd edn (Avignon: Fez, 1757).

Garnerin, A.-J., *Voyage et captivité de [Garnerin], ancien commissaire de la République française, prisonnier d'état en Autriche* (Paris: Librairie du Cercle-Social, 1797).

—, *Détails des trois premiers voyages aériens que M. Garnerin a fait en Russie* (Moscow: Typographie de l'Université, 1803).

—, *Parc de Mousseaux. Voyage aérien du Citoyen Garnerin, avec une jeune personne* (Paris: Imprimerie du Cercle-Social, an VI).

—, *Physique: Voyage aérien* (n.p., n.d).

—, *Rapport fait par le C. Garnerin, de son voyage aérien* ([Paris]: Imprimerie-Librairie du Cercle Social, an VI).

Gattey, M., and H., *Souscription pour la construction d'un vaisseau aérien* (n.p., n.d.).

Gellius, A., *The Attic Nights*, ed. J. C. Rolfe, 3 vols (London: William Heinemann, 1927).

[Gerard, L. G.], *Essai sur l'art du vol aérien* (Paris: Duchesne, 1784).

Gerli, A., *Opuscoli di Agostino Gerli* (Parma: Dalla Samperia Reale, 1785).

Gerli, A., G. Gerli and C. G. Gerli, *Maniera di migliorane e dirigere I alloni aerie* (Rome, 1790).

Gonsales, D., *The Man on the Moone: or A Discourse of a Voyage thither* (London: John Norton, 1638).

Guyot, M., *Essai sur la construction des ballons aérostatique et sur la maniere de les diriger* (Paris: Gueffier, 1784).

Guyton de Morveau, L. B., *Description de l'aérostate* (Dijon: Causse, 1784).

Hapdé, J.-B.-A. and H. Chaussier, *Le Parachute, comédie-parade en un acte et en prose* (Paris: Sages, An VI).

Henin, F., *Mémoire sur la direction des aerostats* (Paris: Moreau, An X).

Histoire intéressante d'un nouveau voyage à la lune, et de la descente à Paris d'une jolie dame de cette terre étrangère (Paris: F.-G. Deschamps, 1784).

Hoole, S., *Thoughts on the Farther Improvement of Aerostation* (London: Nicol, 1785).

Inchbald, E., *The Mogul Tale, or The Descent of the Balloon* (London: Powell, 1796).

[Isemberg, Luigi de.] *Il pallone volante della thuillerie* (N.p., 1784).

Jefferson, T., *Papers*, ed. J. P. Boyd, 10 vols (Princeton, NJ: Princeton University Press, 1950–95).

Jeffries, J., *A Narrative of Two Aerial Voyages of Dr. Jeffries with Mons. Blanchard; with Meteorological Observations and Remarks* (1786; New York, 1941).

Johnson, J. H. *Listening in Paris: A Cultural History* (Berkeley, CA: University of California Press, 1996).

Johnson, S., *The History of Rasselas, The Prince of Abissinia*, ed. J. P. Hardy (Oxford: Oxford University Press, 1999).

—, *The Rambler*, 3 vols (London: Thomas Tegg, 1822–6).

Karamzin, N. M., *Letters of a Russian Traveler, 1789–1790*, trans. F. Jonas (New York: Columbia University Press, 1957).

Kermode, F. and A. Kermode. *The Oxford Book of Letters* (Oxford: Oxford University Press, 1995).

Kramp, C., *Geschichte der aerostatic, historich, physich und mathematisch ausgefuhrt*, 2 vols (Strasburg, 1784).

Kratzenstein, C. G., *L'art de naviguer dans l'air* (Copenhave & Leipzig: Faber & Nitschke, 1784).

Kurzgefasste Beschreibung der Aerostatischen Maschinen (Lyon, 1784).

Lancetti, V., *Areostiade: ossia, il mongolfiero, poema*, 2 vols (Milan: Agnello Nobile, 1803).

Launoy and F. Bienvenu, *Instruction sur la nouvelle Machine inventée par MM. Launoy, Naturaliste, & Bienvenu, Machiniste-Physicien* (n.p., n.d.).

*Lettre à M. de *** sur son projet de voyager avec la Sphere Aérostatique de M. de Montgolfier* (Paris: Marchands de Feuilles Volonts, n.d).

Lettre à Mr. M. de Saint-Just, sur le globe aërostatique de MM. Montgolfier (Paris: Mérigot, Royez, 1784).

Lettre à un ami, sur l'utilité des globes volans, de M. de Montgolfier, & sur la possibilité de la prise de Gibralter (Paris: Gueffier, 1783).

[Lévêque, P.], *Déscription de la second expérience aérostatique faite à Nantes, le 6 Septembre 1784* (Nantes: De Brun, [1784]).

Linguet, S.-N.-H. (ed.), *Annales politiques, civiles et littéraires du dix-huitième siècle*, 19 vols (London: 1777–92).

Low, N., *Low's Almanack and Astronomical and Agricultural Register* (Boston: Munroe & Francis, 1819).

Lunardi, V., *A Viagem Aerea do Capitão Vicente Lunardi, por elle escripta* (Lisbon: Simão Thaddeo Ferreira, 1794).

—, *An Account of the First Aerial Voyage in England* (London, 1784).

—, *An Account of Five Aerial Voyages in Scotland* (London, 1786).

Luxero, M. *Portrait de la beauté, et enlèvement de Vênus dans le Vaisseau-volant de M. Blanchard* (Rome and Paris: Chez les Marchands de Nouveautés, 1783).

[Luzarche], *Nouveaux appareils pour la direction des aérostats, ou essai sur cette direction* (Paris: H.-l. Perronneau, 1812).

Lysons, D., *Collectanea: Or, A Collection of Advertisements and Paragraphs from the Newspapers, Relating to Various Subjects*, 5 vols (Strawberry Hill: T. Kirgate, 1840).

Madam Queasy's Air Balloon, In Madame Queasy's Own Way: A Satirical Poem (1807).

Maistre, X. de., *Prospectus de l'expérience aérostatique de Chambéry, publié au nom des premiers souscripteurs* (Chambéry: F. Puthod, 1784).

Marat, J.-P., *Lettres de l'observateur bon-sens, à M. de ***, sur la fatale catastrophe des infortunes Pilatre de Rozier & Romain* (Paris: Méquignon, 1785).

Martyn, T., *Hints of Important Uses, To be Derived from Aerostatic Globes. With A Print of an Aerostatic Globe, and its Appendages. Originally Designed in 1783*, 2nd edn (London: by the Author, 1784).

Meerwein, C. F., *L'art de voler à la manière des oiseaux* (Basle: J.J. Thourneysen, 1784).

Mémoires secrets pour servir à l'histoire de la République des Lettres en France depuis 1762 jusqu'à nos jours, 36 vols (London: John Adamson, 1784–9).

Mercier, L. S., *Mon Bonnet de Nuit.* (ed.) J.-C. Bonnet (Paris: Mercure de France, 1999).

Mercier, L., and R. de la Bretonne, *Paris le jour, Paris la nuit: "Tableau de Paris," "Le nouveau Paris" de Louis Sébastien Mercier & "Les nuits de Paris" de Restif de la Bretonne.*, ed. M. Delon and D. Baruch (Paris: R. Laffont, 1990).

Money, J., *A Short Treatise on the Use of Balloons and Field Observations in Military Operations* (London: C. Roworth, 1803).

Moore, D., *An Account of Count d'Artois and his Friend's Passage to the Moon, In a Flying Machine, called, an Air Balloon* (Litchfield, CT: Collier, 1785).

Most Grand and Magnificent Aerial Experiment and Diversion. [1785].

Le Mouton, le Canard, et le Coq. Fable dialogue (Bruxelles, 1783).

Moyen de diriger le globe aérostatique, nouvellement découvert (Paris: Le Jay, 1784).

Mr. Blanchard and a Lady (1785).

Nougaret, P.-J.-B., *Tableau mouvant de Paris, ou Variétés amusantes*, 3 vols (Paris: Duchesne, 1787).

Olivari, F., *An Essay on Aérostation* (Dublin: George Bonham, 1797).

Ordonnance de Police, qui fait défenses à toutes personnes de quelque qualité & condition qu'elles soient, de fabriquer & faire enlever aucuns ballons (Angers: Mame, 1784).

Ordonnance de Police, qui fait defenses de fabriquer & faire enlever des Ballons (Paris: Pierres, 1784).

Ordonnantie van den Keyser, Raekende de Ballons ofte Locht-Tuygen met Vuer, genoemt Montgolfieres (Brussels: Pauwels, 1786).

Il Pallon volante, commedia di un atto solo (Perugia: Mario Riginaldi, 1784).

Pecchio, L., *A Elise Garnerin, au sujet de son vol aérostatique exécute à Milan le avril 1824* (Milan: Nicolas Bettoni, 1824).

Petit Journal du Palais Royal, ou Affiches, annonces et avis divers (Paris: Caveau, 1789).

Pilâtre de Rozier, J. F., *Première expérience de la Montgolfiére, construite par ordre du roi* (Paris: L'Imprimierie de Monsieur, 1784).

Pilon, F., *Aerostation; or, The Templar's Stratagem* (London: Kearsley, 1784).

Pingeron, J.-C., *L'art de faire soi-même les ballons aérostatiques, conformes à ceux de M. de Montgolfier* (Paris: Hardouin, 1783).

[Piroux], *L'art de voyager dans les airs, ou les ballons* (Paris, 1784).

Potain, *Relation aérostatique dédiée à la nation Irlandaise* (Paris: Delaunay, 1824).

Réclamation du Mouton, premier navigateur aërien. Paris, Cailleau (1784).

Relation du voyage aérostatique du Général Buonaparte (n.p., n.d.).

Relation d'un voyage aérien à Ching-y-ang dans la tartarie chinoise (London, 1786).

Restif de la Bretonne, N. E., *La Découverte australe*, 2 vols (Leipzig, 1781).

—, *Les Nuits de Paris, ou le Spectateur-nocturne*, ed. M. Delon (Paris: Gallimard, 1986).

Le Retour de mon pauvre oncle, ou relation de son voyage dans la lune (Paris: Lejay, 1784).

Richter, J. P., *Journal de bord de l'aéronaute Giannozzo/Des Luftschiffers Giannozzo Seebuch*, ed. and trans. J. Poumet (Paris: Aubier Montaigne, 1975).

Rigby, E., *An Account of Mr. James Deeker's Two Aerial Expeditions From the City of Norwich* (Norwich: John Crouse, 1785).

Rivarol, A. de, 'Lettre à Monsieur le Président de ***. Sur le Globe Airostatique, sur

les Têtes parlantes, & sur l'état présent de l'opinion publique à Paris', in *Oeuvres*, 5 vols (Paris, 1808; rpt. Geneva: Slatkine, 1968), vol. 2, pp. 207–46.

Robert, A.-Jean and N.-L. Robert. *Mémoire sur les Expériences Aérostatiques* (Paris: L'Imprimerie de P.-D. Pierres, 1784).

Robertson, E.-G., *Mémoires récreatifs, scientifiques et anecdotiques*, 2 vols (Paris: Chez l'auteur, 1831–3).

Rousseau, J.-J., *Le Nouveau Dédale* (Pasedena: The Institute of Aeronautical History, 1950).

Sadler, J., *Balloon: An Authentic Account of the Aerial Voyage of Messrs. Sadler and Clayfield, Who Ascended in a Most Magnificent Balloon From a Field in the Neighbourhood of Stoke's-Croft, Bristol, On Monday, September 24th, 1810*, 2nd edn (Bristol, 1810).

—, *Balloon: An Authentic Narrative of the Aerial Voyage of Mr. Sadler, Across the Irish Channel* (Dublin, 1812).

Saint-Croix, B. Lallemand de. *Procès-verbal très intéressant du voyage aérien qui a eu lieu aux Champs-Elysées le 18 septembre 1791, jour de la proclamation de la constitution* (Paris: De l'Imprimerie du patriote françois, 1791).

Salle, M., *Moyen de diriger l'aérostat* (Paris: Couturier, 1784).

Scott, B., *Aérostat dirigeable à volonté* (Paris: Maradan, 1789).

Ségur, L. P. de. *Memoirs and Recollections* (London: H. Colburn, 1825).

Selle de Beauchamp, Baron de, *Souvenirs de la fin du dix-huitième siècle, extraits des mémoires d'un officier des aérostiers aux armies de 1793–1799* (Paris: Ledoyen et Paul Giret, 1853).

Seward, A., *Letters of Anna Seward*, ed. A. Constable, 6 vols (Edinburgh: Archibald Constable, 1811).

Shelley, P., *The Esdaile Notebook: A Volume of Early Poems*, ed. K. N. Cameron (New York: Knopf, 1964).

Sheridan, B., *Betsy Sheridan's Journal*, ed. W. LeFanu (Oxford: Oxford University Press, 1986).

Suleau, F., *Voyage en l'air: second réveil.* [Paris, 1791].

Supplément à l'art de voyager dans les airs, contenant le Précis historique de la grande Expérience faite à Lyon le 19 Janvier 1784 et l'Exposé d'un moyen ingénieux pour diriger a volonté les Ballons aérostatiques (n.p., n.d.).

Tetu de Brissy, M., *Sur l'exposition de l'Expérience aérostatique faite le 18 juin 1786* (Paris, 1786).

Verhagen, 'Les premiers ballons en Belgique, 1783–1786', *Annales de la Société royale d'archéologie de Bruxelles*, 13 (1899).

Verninac de Saint-Maur. *Epitre à Monsieur Blanchard* (Bruxelles, 1782).

Villiers du Terrage, E. de, *Journal et souvenirs sur l'expédition d'Egypte (1798–1801)* (Paris: Plon, 1899).

Voltaire, F. M. A. de. 'Micromegas', in *Candide and Other Stories*, trans. R. Pearson (Oxford: Oxford University Press, 1990), pp. 101–21.

Watt, J. and J. Black, *Partners in Science: Letters of James Watt and Joseph Black*, ed. E. Robinson and D. McKie (Cambridge, MA: Harvard University Press, 1970).

Wieland, C. M., 'Die Aeropetomanie' and 'Die Aeronauten', in *Sammtliche Werke*, 42 vols (Leipzig: G.J. Göschen, 1794–1801), pp. 30, pp. 1-136.

[Zambeccari, Francesco]. *Proposal for Making a Large Aerostatic Globe* (1784).

—, *Relazione dell'esperienza aereo-statica* (Bologna, 1804).

Secondary Sources

Abate, R., *Storia della aeronautica italiana* (Milan: Bietti, 1974).

Alexander, J., 'Aeromania, "Fire-Balloons," and Catherine the Great's Ban of 1784', *Historian*, 58 (1996), pp. 497–516.

Allan, D., 'Eighteenth-Century Private Subscription Libraries and Provincial Urban Culture: The Amicable Society of Lancaster, 1769–*c.* 1820', *Library History*, 17 (2001), pp. 57–76.

Ambrosini, R., *L'Aereonautica a Bologna, appunti di cronica raccolti* (Bologna: Tipografia di Paolo Neri, 1912).

Anglade, J., *Les Montgolfiers* (Paris: Perrin, 1990).

Aoki, Y., 'To be a Member of the Leading Gentry: The Suffolk Voluntary Subscriptions of 1782', *Historical Research*, 76 (February 2003), pp. 78–92.

Aravamudan, S., *Tropicopolitans: Colonialism and Agency, 1688–1804* (Durham, NC: Duke University Press, 1999).

Arecco, D., *Mongolfiere, science et lumi nel tardo settecento* (Bari: Cacucci, 2003).

Ariotti, P. E., 'Christiaan Huygens: Aviation Pioneer Extraordinary', *Annals of Science*, 36 (1979), pp. 611–4.

Baudrillard, J., *The Consumer Society: Myths and Structures* (London: Sage, 1998).

Beer, M., 'Flying High and Living Large: Eighteenth-Century Balloonists and Twenty-First Century Pop Stars' (Unpublished paper presented at the Spring Annual Research Conference, Agnes Scott College, Decatur, GA, April 2004).

Benedict, B.M., *Curiosity: A Cultural History of Early Modern Inquiry* (Chicago: University of Chicago Press, 2001).

Berg, M., and E. Eger (eds), *Luxury in the Eighteenth-Century: Debates, Desires and Delectable Goods* (New York: Palgrave, 2003).

Berry, A., '"Balancing the Books": Funding Provincial Hospitals in Eighteenth-Century England', *Accounting, Business and Financial History*, 7 (1997), pp. 1–30.

Bianchi, M., 'Taste for Novelty and Novel Tastes: The Role of Human Agency in Consumption', in M. Bianchi (ed.), *The Active Consumer: Novelty and Surprise in*

Boffito, G., *Biblioteca aeronautica italiana* (1929; Firenze: L.S. Olschki, 2002).

—, *Il volo in Italia: Storia documentata e aneddotica dell'aeronautica e dell'aviazione in Italia* (Firenze: G. Barbera, 1921).

Bourdieu, P., *Distinction: A Social Critique of the Judgement of Taste*, trans. R. Nice (Cambridge, MA: Harvard University Press, 1984).

—, *Outline of a Theory of Practice*, trans. R. Nice (Cambridge: Cambridge University Press, 1977).

Boyé, P., *Les premières expériences aérostatiques faites en Lorraine (1783–1788)* (1909; Lacour: Nîmes, 1996).

Bret, P., 'Un bateleur de la science: Le "machiniste-physicien" François Bienvenu et la diffusion de Franklin et Lavoisier', *Annales historiques de la Révolution française*, 338:4 (2004), pp. 95–127.

—, 'Nicolas-Jacques Conté ou l'intelligence technique', *La Revue: Musée des arts et métiers*, 25 (December 1998), pp. 4–13.

—, 'Recherche scientifique, innovation technique et conception tactique d'une arme nouvelle: l'aérostation militaire (1793-1799)', in J.-P. Charnay, et al. (ed.), *Lazare Carnot, ou le savant-citoyen* (Paris: Presses de l'Université de Paris, 1990), pp. 429–51.

Brewer, J. and R. Porter (eds), *Consumption and the World of Goods* (London: Routledge, 1993).

Brewer, J., and F. Trentmann (eds), *Consuming Cultures, Global Perspectives: Historical Trajectories, Transnational Exchanges* (Oxford: Berg, 2006).

Brown, B. T., 'French Scientific Innovation in Late-Eighteenth-Century Dublin: The Hydrogen Balloon Experiments of Richard Crosbie (1783–1785)', in G. Gargett and G. Sheridan (eds), *Ireland and the French Enlightenment, 1700-1800* (New York: St. Martin's, 1999), pp. 107–26.

Byrne, L., *History of Aviation in Ireland* (Tallagh, Ireland: Folens, 1980).

Campbell, C., *The Romantic Ethic and the Spirit of Modern Consumerism* (London: Blackwell, 1987).

—, 'The Sociology of Consumption', in D. Miller (ed.), *Acknowledging Consumption: A Review of New Studies* (London: Routledge, 1995), pp. 96–126.

Canby, C., 'James Sadler, Aeronaut', *British Heritage*, 1:4 (1980), pp. 18–29.

Caproni, T. G., A. Bertanelli and G. Boffito, *Francesco Zambeccari,*

aeronauta (Bologna, 1752–1812) (Milan: Museo Caproni, 1931).

Carlyle, T., *The French Revolution*, 2 vols (New York: Collier, 1900).

Caron, P. and C. Gével, 'Mademoiselle Elisa Garnerin, aéronaute', *Revue politique et littéraire: Revue bleue*, 14 (8 Avril 1912), pp. 434–9.

Cazenove, R. de, 'scension du Ballon le "Gustave' à Lyon, 4 Juin 1784"', *Revue Lyonnaise*, 7 (1884), pp. 549–69.

—, *Premiers voyages aériens à Lyon en 1784* (Lyon: Pitrar, 1887).

Chabaud, G., 'Entre sciences et sociabilités: les experiences de l'illusion artificielle en France à la fin du XVIIIe siècle', *Bulletin de la Société d'Histoire Moderne et Contemporaine*, 44 (1997), pp. 36–44.

—, 'La physique amusante et les jeux expérimantaux en France au XVIIIe siècle', *Ludica*, 2 (1996), pp. 61–73.

Chartier, R., *The Cultural Uses of Print in Early Modern France*, trans. L. G. Cochrane (Princeton, NJ: Princeton University Press, 1987).

—, *Cultural History: Between Practices and Representations*, trans. L. G. Cochrane (Ithaca, NY: Cornell University Press, 1988).

—, *The Cultural Origins of the French Revolution*, trans. L. G. Cochrane (Durham, NC: Duke University Press, 1991).

Chiles, J. R., *The God Machine: From Boomerangs to Blackhawks: The Story of the Helicopter* (New York: Bantam, 2007).

Christopher, J., *Riding the Jetstream: The Story of Ballooning from Montgolfier to Breitling* (London: John Murray, 2001).

Clark, W., J. Golinski and S. Schaffer (eds), *The Sciences in Enlightened Europe* (Chicago, IL: University of Chicago Press, 1999).

Claveran, V. G., 'Globos aerostáticos en la Oaxaca del siglo XVIII', *Quipu*, 4 (1987), pp. 387–400.

Clement, P.-L., *Les Montgolfiers: Leur invention, leur evolution du XVIIIe à nos jours* (Paris: Tardy, 1982).

Cobban, A., *In Search of Humanity: The Role of the Enlightenment in Modern History* (New York: George Braziller, 1960).

Cochrane, E., *Tradition and Enlightenment in the Tuscan Academies, 1690–1800* (Chicago, IL: University of Chicago Press, 1961).

Coley, A., 'Followers of Daedalus: Science and Other Influences in the Tales of Flight in Eighteenth-Century French Literature', *Studies on Voltaire and the Eighteenth Century*, 371 (1999), pp. 81–173.

—, 'The Theme of Experiment in French Literature of the Earlier Eighteenth Century (1715–1761)', *Studies on Voltaire and the Eighteenth Century*, 332 (1995), pp. 213–333.

Connaughton, M. E., '"Ballomania," the American Philosophical Society and Eighteenth-Century Science', *Journal of American Culture*, 7 (1984), pp. 71–4.

Coquery, Natacha. 'French Court Society and Advertising Art: The Reputation of Parisian Merchants at the End of the Eighteenth Century', in C. Wischermann and E. Shore (eds), *Advertising and the European City: Historical Perspectives*, (Aldershot: Ashgate, 2000), pp. 96–112.

Cornu-Thenard, F., *Nicolas-Jacques Conté: Curieux autodidacte du XVIIIe siècle* (Dijon: Jobard, 1955).

Costa, Shelley. "Marketing Mathematics in Early Eighteenth-Century England: Henry Beighton, Certainty, and the Public Sphere." *History of Science* 40 (2002): 211-32.

Coutil, L., *Jean-Pierre Blanchard, physicien-aéronaute* (Evreux: Charles Herissey, 1911).

Crouch, T. D., *The Eagle Aloft: Two Centuries of the Balloon in America* (Washington, DC: Smithsonian Institution, 1983).

Cunningham, A., and N. Jardine (eds), *Romanticism and the Sciences* (Cambridge: Cambridge University Press, 1990).

Darnton, R., *The Business of Enlightenment: A Publishing History of the Encyclopédie, 1775–1800* (Cambridge, MA: The Belknap Press, 1979).

—, *The Literary Underground of the Old Regime* (Cambridge, MA: Harvard University Press, 1982).

—, *Mesmerism and the End of the Enlightenment in France* (Cambridge, MA: Harvard University Press, 1968).

Delon, M. and J. M. Goulemot (eds), *Cahiers de literature française*, vol. 5, *Ballons et regards d'en haut* (Paris: L'Harmattan; Bergame: Bergamo University Press, 2007).

Depuis-Delcourt, J.-F., *Des Ballons dans les fêtes publiques* (Imprimerie: Schneider et Langrand, 1846).

—, *Nouveau manuel complet d'aérostation, ou Guide pour server á l'histoire et á la pratique des ballons* (1850; Paris: Chez Leonce Laget, 1978).

De Vries, J., *The Industrious Revolution: Consumer Behavior and the Household Economy, 1650 to the Present* (Cambridge: Cambridge University Press, 2008).

Digonnet, F., *L'Invention de l'aérostation à Avignon en 1782 et les premières ascensions dans cette ville* (Avignon: Seguin, 1906).

Dollfus, A., *Pilâtre de Rozier: premier navigateur aérien, première victime de l'air* (Paris: Association Française pour l'Avancement des Sciences, 1993).

Douglas, M. and B. Isherwood, *The World of Goods: Towards an Anthropology of Consumption* (London: Routledge, 1996).

Duhem, J., *Histoire de l'arme aérienne avant le moteur* (Paris: Nouvelles editions latines, 1964).

—, *Histoire des idées aéronautiques avant Montgolfier* (Paris: Fernand Sorlot, 1943).

—, *Musée aéronautique avant Montgolfier* (Paris: Sorlot, 1943).

Faure, M., *Les frères Montgolfier et la conquête de l'air* (La Calade, Aix-en-Provence: Edisud, 1983).

Feldhaus, F. M., *Luftfahrten einst und jetzt* (Berlin: H. Paetel, 1908).

Fergusson, J., *Balloon Tytler* (London: Faber and Faber, 1972).

Ferrero, Giovanni, 'Sophie Blanchard, Amazzone del Cielo, in Val Trebbia', *Studi genuensi*, 14 (1998), pp. 21–49.

Feyel, G., 'Négoce et presse provinciale en France au 18e siècle: methods et perspectives de recherches', in F. Angiolini and D. Roche (eds), *Cultures et formations négociantes dans l'Europe moderne* (Paris: Editions de l'Ecole des Hautes Etudes en Sciences Sociales, 1995), pp. 439–511.

Fontaine, R., *La Manche en ballon: Blanchard contre Pilâtre de Rozier* (Dunkerque: Westhoek-Editions, 1982).

Fox, R., 'Science and Government', in *The Cambridge History of Science*, vol. 4., *Eighteenth-Century Science*, ed. R. Porter (Cambridge: Cambridge University Press, 2003), pp. 107–28.

Freeman, D., 'Riding the Wind in Charliers and Montgolfiers', *Early American Life*, 8:4 (1977), pp. 34–7, 64–5.

Freska, F. (ed.), *A Peace Conference of Intrigue*, trans. H. Hansen (New York: The Century Co., 1919).

Galitzyne, N., 'Les premières expériences de Montgolfier: D'après des documents russes', in *Annales internationals d'histoire. Congrès de Paris, 1900*, 7 vols in 2 tomes (Nendeln/Liechtenstein: Kraus Reprint, 1972), tome 2, vol. 5, pp. 146–54.

Gardiner, L., *Man in the Clouds: The Story of Vincenzo Lunardi* (Edinburgh: Chambers, 1963).

Gay, P., *The Enlightenment: An Interpretation*, 2 vols (New York: Norton, 1966–9).

Gillespie, R., "Ballooning in France and Britain, 1783–1786", *Isis*, 75 (1984), pp. 249–68.

Gillispie, C. C., *The Montgolfier Brothers and the Invention of Aviation: 1783-1784* (Princeton, NJ: Princeton University Press, 1983).

—, *Science and Polity in France: The Revolutionary and Napoleonic Years* (Princeton: Princeton University Press, 2004).

Glennie, P., 'Consumption within Historical Studies', in D. Miller (ed.), *Acknowledging Consumption: A Review of New Studies* (London: Routledge, 1995), pp. 164–203.

Glines, C. V., 'First in America's Skies', *Aviation History*, 7 (1996), pp. 26–32.

Godechot, J., 'L'Aérostation militaire sous le directoire', *Annales historiques de la Révolution française*, 8 (1931), pp. 213–28.

—, 'Les premiers soldats du ciel: Débuts de l'aérostation militaire', Historama, 11 (1985), pp. 24–8.

Golinski, J., 'Barometers of Change: Meteorological Instruments as Machines of Enlightenment', in Clark, Golinski and Schaffer (eds), *The Sciences in Enlightened Europe*, (Chicago, IL: University of Chicago Press, 1999), pp. 69–93.

Guth, P., 'Tous les arts se sont mis "au ballon"', *Connaissance des Arts*, 59 (1957), pp. 14–19.

Hargreaves-Mawdsley, W. N., *The English Della Cruscans and Their Time, 1783–1828* (The Hague, Martinus Nijhoff, 1967).

Heggen, Alfred. 'Die "ars volandi" in der Literatur des 17 und 18 Jahrhunderts', *Technikgeschichte*, 42 (1975), pp. 327–37.

Hilaire-Pérez, L., *L'invention technique au siècle des Lumières* (Paris: Albin Michel, 2000).

Hill, R. D., 'The Search for Peter Carnes', *Richmond County History*, 10:2 (1978), pp. 5–10.

Hodgson, J. E., *Doctor Johnson on Ballooning and Flight* (London: Elkin Mathews, 1925).

—, *The History of Aeronautics in Great Britain* (London: Oxford University Press, 1924).

Holl, P., 'Les débuts de l'aviation à Strasbourg', *La Vie en Alsace* (1926), pp. 137–42.

Horne, Alistaire. "The Balloons of Paris." *MHQ: The Quarterly Journal of Military History*, 13 (2001), pp. 80–7.

Hunn, J. M., 'The Balloon Craze in France, 1783–1799: A Study in Popular Science', (Ph.D. dissertation, Vanderbilt University, 1982).

Jones, C., 'The Great Chain of Buying: Medical Advertisement, the Bourgeois Public Sphere and the Origins of the French Revolution', *American Historical Review*, 101 (1996), pp. 13–40.

Jones, J., *Sexing la Mode: Gender, Fashion and Commercial Culture in Old Regime France* (Oxford : Berg, 2004).

Keen, P., 'The "Balloonomania": Science and Spectacle in 1780s England', *Eighteenth-Century Studies*, 39 (2006), pp. 507–35.

Kerr, Joan Patterson. "'What Good is a New-Born Baby?' Of Balloons, the First Air-Mail Letters, and the Ever-Enterprising Franklin Family." *American Heritage* 25:1 (1973): 24-31.

Kim, Mi G., ""Public" Science: Hydrogen Balloons and Lavoisier's Decomposition of Water', *Annals of Science*, 63 (2006), pp. 291–318.

Korzus, B., and B. Leismann (eds), *Leichter als Luft: Zur Geschichte der Ballonfahrt* (Münster: Landschaftverband Westfalen-Lippe, 1978).

Kraft, E., 'Anna Letitia Barbauld's "Washing Day" and the Montgolfier Balloon', *Literature and History*, 4:2 (1995), pp. 25–41.

Kuhn, G., 'Fiestas and Fiascoes – Balloon Flights in Nineteenth-Century Mexico', *Journal of Sport History*, 13:2 (1986), pp. 111–18.

—, 'The History of Aeronautics in Latin America' (Ph.D. dissertation, University of Minnesota, 1965).

Kwass, M., 'Ordering the World of Goods: Consumer Revolution and the Classification of Objects in Eighteenth-Century France', *Representations*, 82 (2003), pp. 87–116.

—, 'Big Hair: A Wig History of Consumption in Eighteenth-Century France,' *American Historical Review*, 111 (2006), pp. 631–59.

Lacroix, D., *Les aérostiers militaires du Chateau de Meudon (1794–1884)* (Paris: Auguste Ghio, 1885).

Landa, L.A., 'Johnson's Feathered Man: "A Dissertation on the Art of Flying" Considered," in W.H. Bond (ed.), *Eighteenth-Century Studies in Honor of Donald F. Hyde* (New York: The Grolier Club, 1970), pp. 161–78.

Langins, J., 'Hydrogen Production for Ballooning during the French Revolution: An Early Example of Chemical Process Development', *Annals of Science*, 40 (1983), pp. 531–58.

Lecornu, Joseph. *La navigation aérienne: Histoire documentaire et anecdotique*, 3rd edn (Paris: Vuibert & Nony, 1910).

Leroudier, E., *Noëls et chansons en patois Lyonnais du XVIIIe siècle* (Lyon: Cumin et Masson, 1918).

Lough, J., *France on the Eve of Revolution: British Travellers' Observations* (London: Croom Helm, 1987).

Lunney, L., 'The Celebrated Mr. Dinwiddie: An Eighteenth-Century Scientist in Ireland', *Eighteenth-Century Ireland*, 3 (1988), pp. 69–83.

Lynn, M. R. *Popular Science and Public Opinion in Eighteenth-Century France* (Manchester: Manchester University Press, 2006).

—, 'Sparks for Sale: The Culture and Commerce of Fireworks in Early Modern France', *Eighteenth-Century Life*, 30 (2006), pp. 74–97.

MacCarron, Donal. *A View from Above* (Dublin: The O'Brien Press, 2000).

Mandelbaum, N. B., 'Jean-Paul Marat: The Rebel as Savant (1743–1788). A Case Study in Careers and Ideas at the End of the Enlightenment', (Ph.D. Dissertation, Columbia University, 1977).

Mason, M., *Aeronautica; or, Sketches Illustrative of the Theory and Practice of Aerostation* (London: F.C. Westley, 1838).

McCarthy, W., *Anna Letitia Barbauld: Voice of the Enlightenment* (Baltimore, MD: The Johns Hopkins University Press, 2008).

McClellan, J. E., *Colonialism and Science: Saint Dominque in the Old Regime* (Baltimore, MD: The Johns Hopkins University Press, 1992).

McCloy, S.T., *French Inventions of the Eighteenth Century* (Louisville, KY: University of Kentucky Press, 1952).

McCracken, G., *Culture and Consumption: New Approaches to the Symbolic Character of Consumer Goods and Activities* (Bloomington: Indiana University Press, 1990).

McKendrick, N., J. Brewer and J. H. Plumb, *The Birth of a Consumer Society: The Commercialization of Eighteenth-Century England* (Bloomington, IN: Indiana University Press, 1982).

Melton, J. V. H., *The Rise of the Public in Enlightenment Europe* (Cambridge: Cambridge University Press, 2001).

Miller, D. (ed.), *Acknowledging Consumption: A Review of New Studies* (London: Routledge, 1995).

—, *Material Culture and Mass Consumption* (London: Blackwell, 1987).

Monjau, H., *Wilhelmine Reichard: erste deutsche Ballonfahrerin: 1788 bis 1848: eine dokumentarische Biographie* (Freital: Monjau, 1998).

Morazzoni, G., *Un pioniere dell'aereonautica, Vincenzo Lunardi* (Milan: W. Toscanini, 1931).

Mukerji, C., *From Graven Images: Patterns of Modern Materialism* (New York: Columbia University Press, 1983).

Nicolson, Majorie Hope. *Science and Imagination* (Ithaca: Great Seal Books, 1956).

—, *Voyages to the Moon* (New York: Macmillan, 1949).

—, *A World in the Moon: A Study of the Changing Attitude Toward the Moon in the Seventeenth and Eighteenth Centuries* (Northhampton, MA: Smith College, 1936).

Norton, M. B., 'America's First Aeronaut: Dr. John Jeffries', *History Today*, 18 (1968), pp. 722–9.

Oberling, P., 'A History of Turkish Aviation. Part I: Aerostation among the Ottomans', *Archivum Ottomanicumi*, 9 (1984), pp. 129–62.

Oettermann, S., 'Die Fliegenden Plastiken des Johann Karl Enslen: Johann Karl Enslen's Flying Sculptures', *Daidalos: Berlin Architectural Journal*, 37 (15 September 1990), pp. 44–53.

Outram, D., *The Enlightenment* (Cambridge: Cambridge University Press, 1995).

Penny, J., *Up, Up and Away: An Account of Ballooning in and around Bristol and Bath, 1784–1999* (Bristol: Bristol Branch of the Historical Association, 1999).

Perrin, C., *La vie rocambolesque d'André Garnerin, Pionnier du parachute* (Paris: Messene, 2000).

Poniatowski, M., *Garnerin, le premier parachutiste de l'histoire* (Paris: Albin Michel, 1983).

Poterlet, M., *Notice sur Madame Blanchard, aéronaute* (Paris: Imprimerie de Fain, 1819).

Queruel, A., *Nicolas-Jacques Conté (1755–1805), un inventeur de génie: Des crayons à l'Expédition d'Egypte en passant par l'aérostation militaire* (Paris: L'Harmattan, 2004).

Quinlan, M., 'Balloons and the Awareness of a New Age', *Studies in Burke and His Time*, 14 (1972–3), pp. 221–38.

Ravel, J., *The Contested Parterre: Public Theater and French Political Culture, 1680–1791* (Ithaca, NY: Cornell University Press, 1999).

Riskin, J., *Science in the Age of Sensibility: The Sentimental Empiricists of the French Enlightenment* (Chicago, IL: University of Chicago Press, 2002).

Robbins, L. E. *Elephant Slaves and Pampered Parrots: Exotic Animals in Eighteenth-Century Paris* (Baltimore, MD: Johns Hopkins University Press, 2002).

Robinson, M., *The Coldest Crucible: Arctic Exploration and American Culture* (Chicago, IL: University of Chicago Press, 2006).

Roberts, L., 'Going Dutch: Situating Science in the Dutch Enlightenment', in W. Clark, J. Golinski and S. Schaffer (eds), *The Sciences in Enlightened Europe*, (Chicago, IL: University of Chicago Press, 1999), pp. 350–88.

Robineau, L., 'Lazare Carnot et les Compagnies d'aérostiers', *Revue historique des Armées*, 2 (1989), pp. 101–10.

Rocamora, M., *Historia de la navegación aérea en Barcelona* (Barcelona: José Porter, 1948).

Roche, D., *A History of Everyday Things: The Birth of Consumption in France, 1600–1800*, trans. B. Pearce (Cambridge: Cambridge University Press, 2000).

Rolt, L. T. C., *The Aeronauts: A History of Ballooning, 1783–1903* (New York: Walker, 1966).

Rosenau, H., 'The Sphere as an Element in the Montgolfier Monuments', *Art Bulletin*, 50 (1968), pp. 65–6 and plates.

Rousseau, G. S., and R. Porter (eds), *Exoticism and the Enlightenment* (Manchester: Manchester University Press, 1990).

Schaffer, S., 'The Consuming Flame: Electrical Showmen and Tory Mystics in the World of Goods', in J. Brewer and R. Porter (eds), *Consumption and the World of Goods* (London: Routledge, 1993), pp. 489–526.

Schama, S., *Citizens: A Chronicle of the French Revolution* (New York: Vintage, 1989).

Schneider, R. R., 'Star Balloonist of Europe: The Career of Marie-Madeleine Blanchard', *Consortium on Revolutionary Europe, 1750–1850* (1983), pp. 696–711.

Schofield, R. E. *The Lunar Society of Birmingham: A Social History of Provincial Science and Industry in Eighteenth-Century England* (Oxford: Oxford University Press, 1963).

Schwartz, V. R., *Spectacular Realities: Early Mass Culture in Fin-de-Siècle Paris* (Berkeley: University of California Press, 1998).

Secord, J., 'Newton in the Nursery: Tom Telescope and the Philosophy of Tops and Balls, 1761–1838', *History of Science*, 23 (1985), pp. 127–51.

Sgard, J., 'Les philosophes en montgolfière', *Studies on Voltaire and the Eighteenth Century*, 303 (1992), pp. 99–111.

—, (ed.), *La presse provinciale au XVIIIe siècle* (Grenoble: Centre de Recherches sur les Sensibilités, 1983).

Shapin, S., 'The Audience for Science in Eighteenth-Century Edinburgh', *History of Science*, 12 (1974), pp. 95–121.

Simmel, G., 'Fashion', *International Quarterly*, 10 (October 1904), pp. 130–55.

Smith, W.D., *Consumption and the Making of Respectability*, 1600–1800 (London:Routledge, 2002).

Straub, H., *Fliegen mit Feuer und Gas: Die Geschichte der Ballon- und Luftschiffahrt* (Aarau: AT Verlag, 1984).

Stevenson, D., *The Beggar's Benison: Sex Clubs of Enlightenment Scotland and their Rituals* (East Linton, East Lothian, Scotland: Tuckwell Press, 2001).

Stewart, L., *The Rise of Public Science: Rhetoric, Technology, and Natural Philosophy in Newtonian Britain, 1660-1750* (Cambridge: Cambridge University Press, 1992).

Styles, J., 'Product Innovation in Early Modern London', *Past and Present*, 168 (2000), pp. 124–69.

Swift, J., *Gulliver's Travels* (Indianapolis, IN: Bobbs-Merrill, 1963).

Targosz, K., '"Le dragon volant" de Tito Livio Burattini', *Annali dell'Istituto e Museo di Storia della Scienza di Firenze*, 2:2 (1977), pp. 67–85.

Taton, R., 'Nouveau document sur le "dragon volant" de Burattini', *Annali dell'Istituto e Museo di Storia della Scienza di Firenze*, 7:2 (1982), pp. 161–8.

Le Temps des Ballons: Art et Histoire (Paris: Editions de La Martinière, 1994).

Thébaud-Sorger, M., *L'Aérostation au temps des Lumières* (Rennes: Presses universitaires de Rennes, 2009).

—, 'Innovation and Risk Management in Late Eighteenth-Century France: The Administration of Inventions in French Cities at the End of the Old Regime', in C. Rabier (ed.), *Fields of Expertise: A Comparative History of Expert Procedures in Paris and London, 1600 to Present* (Newcastle: Cambridge Scholars Publishing, 2007), pp. 261–89.

Thompson, E. P., 'The Moral Economy of the English Crowd in the Eighteenth Century', *Past and Present*, 50 (1971), pp. 76–136.

Tissandier, G., *Bibliographie aeronautique* (Paris: Launette, 1887).

—, *Histoire des ballons et des aeronauts célèbres* (Paris: Launette, 1887).

Todd, C., 'French Advertising in the Eighteenth Century', *Studies on Voltaire and the Eighteenth Century*, 266 (1989), pp. 513–47.

Turner, K., 'The Spectacle of Democracy in the Balloon Plays of the Revolutionary Period', *Forum for Modern Language Studies*, 39 (2003), pp. 241–53.

Turner, V., *From Ritual to Theater: The Human Seriousness of Play* (New York: Performing Arts Journal Publications, 1982).

Turnor, C. H., *Astra Castra: Experiments and Adventures in the Atmosphere* (London: Chapman and Hall, 1865).

Vallès I Rovira, I., *La Màgia del Vol: Primeres Proves Aerostàtiques a Barcelona, València I Castella al Final des Segle XVIII* (Barcelona: Alta Fulla, 1985).

Venturi, F., *The End of the Old Regime in Europe, 1776-1789*, trans. R. Burr Litchfield, 2 vols (Princeton: Princeton University Press, 1991).

Verhaegen, P. 'Les premiers ballons en Belgique, 1783–1786', *Annales de la Société royale d'archéologie de Bruxelles*, 13 (1899), pp. 361–72.

Vila, A., *Enlightenment and Pathology: Sensibility in the Literature and Medicine of Eighteenth-Century France* (Baltimore, MD: Johns Hopkins University Press, 1998).

Vindel, P., *Historia bibliogáfica e iconográfica de la aeronáutica en España, Portugal, Países Hispano-Americanos y Filipinas desde lor origenes hasta 1900* (Madrid: Pedro Vindel, 1930).

Wallner, A., *The First Air Voyage in the United States: The Story of Jean-Pierre Blanchard* (New York: Holiday House, 1996).

Walsh, C., 'The Advertising and Marketing of Consumer Goods in Eighteenth-Century London', in Wischermann and Shore (eds), *Advertising and the European City: Historical Perspectives* (Aldershot: Ashgate, 2000), pp. 79–95.

Walters, A., 'Conversation Pieces: Science and Politeness in Eighteenth-Century England', *History of Science*, 35 (1997), pp. 121–54.

—, 'Ephemeral Events: English Broadsides of Early Eighteenth-Century Solar Eclipses', *History of Science*, 37 (1999), pp. 1–43.

Waquet, F., 'Book Subscriptions in Early Eighteenth-Century Italy', *Publishing History*, 33 (1993), pp. 77–88.

Weatherill, L., 'The Meaning of Consumer Behaviour in Late Seventeenth- and Early Eighteenth-Century England', in J. Brewer and R. Porter (eds), *Consumption and the World of Goods*, (London: Routledge, 1993), pp. 206–27.

Weber, M., *The Protestant Ethic and the Spirit of Modern Capitalism*, trans. T. Parsons (New York: Scribner's, 1958).

Welch, E., *Shopping in the Renaissance: Consumer Cultures in Italy, 1400-1600* (New Haven, CT: Yale University Press, 2005).

Werrett, S., 'From the Grand Whim to the Gasworks: "Philosophical Fireworks" in Georgian England', in *The Mindful Hand: Inquiry and Invention From the Late Renaissance to Early Industrialisation*, ed. L. Roberts, S. Schaffer and P. Dear (Amsterdam: Koninklijke Nederlandse Akademie van Wetenschappen, 2007), pp. 325–47.

Williams, L. P., 'The Politics of Science in the French Revolution', in M. Clagett (ed.), *Critical Problems in the History of Science* (Madison, WI: University of Wisconsin Press, 1959), pp. 291–308.

Wischermann, C., 'Placing Advertising in the Modern Cultural History of the City', in C. Wischermann and E. Shore, *Advertising and the European City: Historical Perspectives* (Aldershot: Ashgate, 2000), pp. 1–31.

Wouwermans, A., *Contribution à la bibliographie de la locomotion aérienne*. Anvers, 1894.

INDEX

For Product Safety Concerns and Information please contact our EU representative GPSR@taylorandfrancis.com
Taylor & Francis Verlag GmbH, Kaufingerstraße 24, 80331 München, Germany

www.ingramcontent.com/pod-product-compliance
Ingram Content Group UK Ltd.
Pitfield, Milton Keynes, MK11 3LW, UK
UKHW011002180926
475701UK00004B/516
* 9 7 8 1 1 3 8 6 6 1 2 1 9 *